SpringerBriefs in Computer Science

For further volumes:
http://www.springer.com/series/10028

Seungyeol Lee

Glazed Panel Construction with Human–Robot Cooperation

Springer

Seungyeol Lee
Robotics Research Division
Daegu Gyeongbuk Institute of Science
and Technology
50-1, Sang-Ri, Hyeonpung-Myeon,
Dalseong-Gun, Daegu 711-813
Republic of Korea
suprasy@paran.com

ISSN 2191-5768 e-ISSN 2191-5776
ISBN 978-1-4614-1417-9 e-ISBN 978-1-4614-1418-6
DOI 10.1007/978-1-4614-1418-6
Springer New York Dordrecht Heidelberg London

Library of Congress Control Number: 2011936141

Cover design: eStudio Calamar, Berlin/Figueres

Printed on acid-free paper

Springer is part of Springer Science+Business Media (www.springer.com)

Preface

Recently, the tendency for buildings and structures is to be ever larger and taller. A new method of construction is, therefore, required to follow the current trend, and construction machinery and equipment are being developed to help in the process at various construction sites. An imbalance in technical manpower supply and demand is a serious problem on construction sites. This labor shortage and the aging of the skilled worker on the construction site cause a lasting increase in wages. These technical service problems have appeared to lower functionality in construction situations. Consequently, the rising labor costs, poorer execution quality, delayed construction period, and increased construction expenses have led to reduced safety and more accidents in construction areas. This problem corresponds to the potential for "automation system and robotics in construction" as one of the solutions. It is possible to substitute an automatic system and robot for technical manpower, which simultaneously increases the working speed and construction quality, improves safety and reduces the cost of construction.

Generally, almost half of construction work is said to be material handling. Materials and equipment used for construction are heavy and bulky for humans. Through the case studies on constructions, to which building material handling robot was applied, however, we could find some factors to be improved. Unlike the automation lines of the general manufacturing industry, construction sites rarely shows repeated operational patterns use to its unstructured processes. That is, construction robots execute orders while operating in a dynamic environment where structures, operators, and equipment are constantly changing. Therefore, a guidance or remote–controlled system is the natural way to implement construction robot manipulators. However, a remote–controlled system has to solve some problems. One of the solutions to address these problems is the technology of human–robot cooperative manipulation.

The purpose of this study is to develop human–robot cooperative manipulation technology to solve all kinds of problems generated by the current installation method, which depends on manpower or a low-quality construction robot for the installation of heavy and bulk building materials at construction sites. The essential technologies of human–robot cooperative manipulation are considered through the analysis of an existing installation method. The prototype's hardware design and control algorithm development are achieved using the results of this analysis. The developed prototype system is corrected and complemented based on the results of a performance test. To apply human–robot cooperative manipulation at real construction sites, we executed additional work required for application. After application to real construction sites, evaluation on the productivity and safety of the developed system was done by comparing and analyzing with the existing installation methods. Lastly, I would like to acknowledge that this book would not exist without the support of my spouse, SAMSUNG CORPORATION.

Notations

Symbol	Description
$\underline{\mathbf{F}}_{\mathbf{h}}(\underline{\mathbf{T}}_{\mathbf{h}})$	The force (torque), measured by the operational force sensor which is generated by the interaction between the operator and HRI device
$\underline{\mathbf{F}}_{\mathbf{e}}(\underline{\mathbf{T}}_{\mathbf{e}})$	The force (torque), measured by the environmental force sensor which is generated by the interaction between the environment and a heavy material
$\mathbf{M}_{\mathbf{pt}}(\mathbf{M}_{\mathbf{ot}})$, $\mathbf{B}_{\mathbf{pt}}(\mathbf{B}_{\mathbf{ot}})$	The impedance parameters, that are related to a desired dynamic behavior, are $M_{pt}(M_{ot})$ and $B_{pt}(B_{ot})$, (n by n positive definite diagonal inertia and damping matrices), 'p' stands for the position and 'o' stands for the orientation
λ	The force augmentation ratio of an operator
$\mathbf{M}_{\mathbf{pe}}(\mathbf{M}_{\mathbf{oe}})$, $\mathbf{B}_{\mathbf{pe}}(\mathbf{B}_{\mathbf{oe}})$, $\mathbf{K}_{\mathbf{pe}}(\mathbf{K}_{\mathbf{oe}})$	The impedance parameters that determine a dynamic behavior of the end effector for interactions with an environment
$\{D\}$	The desired frame specified by a desired position vector $\underline{\mathbf{p}}_{\mathbf{d}}$ and a desired rotation matrix $\mathbf{R}_{\mathbf{d}}$
$\{C\}$	The compliant frame specified by a position vector $\underline{\mathbf{p}}_{\mathbf{c}}$ and a rotation matrix $\mathbf{R}_{\mathbf{c}}$
$\underline{\mathbf{p}}_{\mathbf{e}}$, $\mathbf{R}_{\mathbf{e}}$	The actual end effector position and orientation
$\dot{\underline{\mathbf{p}}}_{e}$, ω_e	The actual end-effector linear velocity and angular velocity
$\mathbf{B}_{\mathbf{h}}$	The damping coefficient of an operator's arm
$\mathbf{K}_{\mathbf{h}}$	The stiffness coefficient of an operator's arm
$\ddot{\underline{\mathbf{X}}}_{d}$	Desired acceleration related target dynamics
$\dot{\underline{\mathbf{X}}}_{d}$	Desired velocity related target dynamics

Contents

1 Introduction 1
1.1 Problems in Construction Industry 1
1.2 Panel Construction Robots 2
1.3 Needs for Human–Robot Cooperative Manipulation 3
References 5

2 Control Algorithm for Human–Robot Cooperation 7
2.1 Outline 7
2.2 System Modeling 8
2.2.1 Operation in Unconstrained Case 8
2.2.2 Operation in Constrained Case 9
2.3 Control Algorithm 11
2.4 Experimental System 12
2.5 Experiment and Results 13
2.5.1 Influence of Impedance Parameters 15
2.5.2 Influence of Inner Motion Control Loop 17
2.5.3 Influence of Force Augmentation Ratio 18
2.5.4 Influence of an Environmental Stiffness Parameter 21
References 22

3 Conceptual Design of Human–Robot Cooperative System 23
3.1 Jab Analysis 23
3.2 System Configuration 24

4 Prototype for Glazed Panel Construction Robot 31
4.1 Basic System 31
4.1.1 6DOF Manipulator 31
4.1.2 Mobile Platform 34

4.2 Additional Module. 36
4.2.1 Hardware 36
4.2.2 Software 38
4.3 Experiments and Results 40
Reference 45

5 Glazed Ceiling Panel Construction Robot 47
5.1 Basic System 47
5.1.1 Multi-DOF Manipulator. 48
5.1.2 Aerial Work Platform 49
5.2 Additional Module. 52
5.2.1 Human Robot Interface 52
5.2.2 Deck Design of Aerial Work Platform. 53
5.2.3 Vacuum Suction Device. 53
5.3 Design of Robotized Construction Process 55
5.4 Field Tests and Results 58
Reference 64

6 Conclusion and Future Works 65

Index 69

Chapter 1
Introduction

Abstract These days, construction companies are beginning to be concerned about a potential labor shortage by demographic changes and an aging construction work force. Also, an improvement in construction safety could not only reduce accidents but also decrease the cost of the construction, and is therefore one of the imperative goals of the construction industry. These challenges correspond to the potential for Automation and Robotics in Construction as one of solutions. Almost half of construction work is said to be material handling and materials used for construction are heavy and bulky for humans. To date, various types of robots have been developed for glazed panel construction. Through the case studies on constructions, to which the robots were applied, however, we could find some difficulties to be overcome. In this study, a human–robot cooperative system is deduced as one approach to surmount these difficulties and then, considerations on interactions among the operator, robot and environment are applied to design of the system controller. The human–robot cooperative system can cope with various and untypical constructing environment through the real-time interacting with a human, robot and constructing environment simultaneously. The physical power of a robot system helps a human to handle heavy construction materials with relatively scaled-down load. Also, a human can feel and response the force reflected from robot end effecter acting with working environment.

Keywords Construction industry · Construction robots · Building materials · Labor shortage · Panel construction · Remote-controlled construction robot · Curtain walls Human–robot cooperative manipulation

1.1 Problems in Construction Industry

Although the construction industry so far has managed to develop highly productive systems without the help of robots, but there are specific areas of application that the industry would benefit from robots application. The main ones are the application of

S. Lee, *Glazed Panel Construction with Human–Robot Cooperation*,
SpringerBriefs in Computer Science, DOI: 10.1007/978-1-4614-1418-6_1,

robots in construction processes which can produce better quality with faster production. Another application area is in hazardous work environment that would be dangerous for a human operator. The robots can be also used to perform construction tasks that are boring or tiring for the human operators [1]. Examples of these construction robots include wall (façade) climbing robots for inspection and maintenance, concrete power floating machines, concrete floor surface finishing robots, construction steel frame welding robots, wall panels' bricklaying robots, robotic excavators and automated cranes for the assembly of modular construction elements [2–8]. Other advances in the automation in construction have been reported on the software side where IT applications have been developed, to increase the safety standards of the construction site, to assist in better planning and execution of projects, to automate the buildings' design process, to visualize the community projects using immersive 3D VR techniques and to monitor and control of the parts and materials flow (through tags and RFID) of the entire construction process.

However, the construction industry is one of the most unfamiliar R&D fields in the robotics and automation community since introducing robotic technology into construction has been more difficult than in other industries. Nevertheless, the construction industry is one of the oldest and largest economic sectors. Construction is a diverse industry characterized by almost unique circumstances for each project and a dynamic unstructured environment, with safety hazards, temporary activities and changing weather conditions, which all together hold back greater automation. In construction automation, the building serves simultaneously as the work environment. The work environment is also constantly exposed to changing weather conditions. To address these conditions unique to construction, the form of construction robots is different from industrial robots [9]. Building materials and components are much larger and heavier than most industrial materials. Buildings are also made of many kinds of materials, and each material may be a different shape. It is evident that the level of automation in construction is very low in comparison with current technological advances.

The objective of early construction robots was to substitute them for skilled labor. It did not take long before researchers discovered that replacing workers is not possible, particularly when it came to recognition and judgment skills. Robots do not have the same capability skilled workers have to inspect their own work in real time and correct any defects. Robots also need instruction, cleaning, preparation, storage and transportation, all of which require the time and effort of a worker [10]. However, the construction industries are being challenged by a serious labor shortage. Today, most of construction companies are beginning to be concerned about a potential labor shortage by demographic changes and an aging construction work force.

1.2 Panel Construction Robots

Almost half of construction work is said to be material handling. Handling heavy materials has been, for the most part, eliminated for outside work by cranes and other various lifting equipment. Such equipment, however, is not available for inside work. Inside work is physically demanding because the use of cranes and other heavy lifting equipment is not practical. Making matters worse, materials, such as plaster board, tend to be heavy and bulky for worker to handle.

However, recent years have seen an increase in the development of construction robot and automated machines that carry out complex sequences of operations in the panel construction sector with great performance. For example, Shimizu's CFR1 is used for ceiling board installation, Taisei's Boardman-100 is used for interior wall board installation and Komatsu's Mighty Hand is used for multi-purpose material handling [10]. The development goals were to relieve workers from hard work and to increase productivity. Common features to all three robots include; each lifts and moves one board at a time to its desired position, the attachment of the board to its substructure must be done by hand and each manipulator has five degrees of freedom (rotating, tilting, side shifting, arm lifting and reaching). Differences among the robots are; handling tasks are programmed by either automatic (sequence) or manual remote control, the power supply required is either 100 or 200 V AC and the end-effector is either a clamp, a vacuum suction cup or an interchangeable attachment.

Although these robots relieve workers form handling heavy materials, it still left difficulties in precision construction and operation of the robots. CFR1's travel course must be determined and input before operation. Once started, it automatically lifts a board to the ceiling, slides into its exact position and holds it while it is manually screwed in place. Unlike CFR1, the goal of Boardman-100 was never to fully automate the installation process. Rather, its only objective was to carry the weight of panels for easier handling. Mighty Hand is controlled remotely from a small control box light enough to carry over a worker's shoulder. A worker must then direct the robot to travel, lift and place materials.

In construction, the product is custom-made and robots must be reprogrammed to operate under each given condition. Consequently, construction robots are defined as field robots that execute orders while operating in a dynamic environment where structures, operators, and equipment are constantly changing. Therefore, a guidance or remote-controlled system is the natural way to implement construction robot manipulators [11, 12]. However, during operation of a remote-controlled construction robot, problems arise due to operators receiving limited accurate information; the contact force applied by the robot can damage building materials such as pit from the contact force, thus reducing the ability to respond to the constantly changing operational environments. In addition, it is difficult to cope with malfunctions immediately when unexpected situations occur during construction work.

1.3 Needs for Human–Robot Cooperative Manipulation

As another case study in a panel construction robot, Samsung sponsored a project to develop a robot to install curtain walls [13, 14]. Aluminum frames and glass are assembled into curtain walls in the factory and conveyed to the construction site

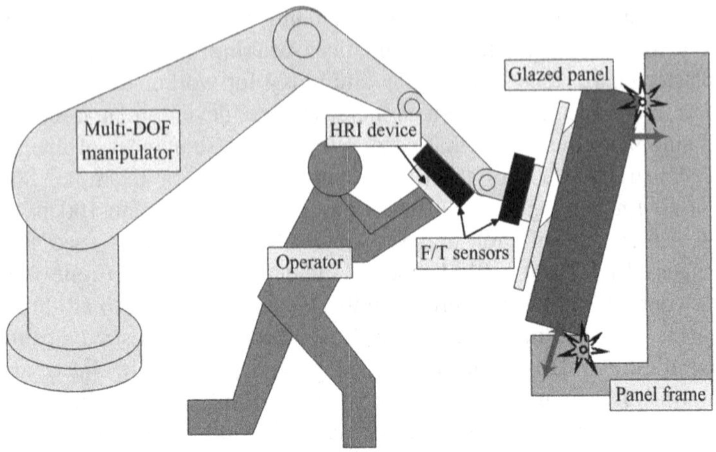

Fig. 1.1 Concept of human–robot cooperative manipulation

for installation. Curtain walls, with an average weight of 400 kg and over 4 m tall, are large and heavy and it is very laborious to handle and install them. However, the developed robot enabled simpler and more precise installation than the existing construction method did, and, most of all, improved safety during installation works. Through the case studies on constructions, to which the robot was applied, however, we could find some factors to be improved. Unlike the automation lines of the general manufacturing industry, construction sites rarely shows repeated operational patterns use to its unstructured processes. Thus, we deduced the following improvements.

- A robot that can follow operator intention in various works at unstructured construction sites.
- A robot that shares work space with an operator.
- Coordination of operator's force and the robot's amplified force.
- Intuitive operational method that can reflect dexterity of an operator.

One of the solutions for these requirements is the technology of human–robot cooperative manipulation. To introduce the human–robot cooperative manipulation, a novel concept of robot manipulation is proposed, as shown in Fig. 1.1, for installation of heavy glazed panels in cooperation between an operator and a robot. Especially, considerations on interactions among the operational force, robot and environment are applied to design of the robot controller. That is to say, the system, to which the introduced control method is applied, allows an operator to handle heavy materials as if he did it by himself or herself, by exerting operational force with a certain force augmentation ratio. Also, this system enables an operator to perform operations more intuitively by allowing him or her to feel reaction forces from environments during an operation.

References

1. Albus, J. S. (1986) Trip report: Japanese progress in robotics for construction. Robotics Magazine, 51:103–112
2. Gambao, E., Balaguer, C., Gebhart, F. (2000) Robot assembly system for computer-integrated construction. Automation in Construction, 9(5/6):479–487
3. Choi, H. S., Han, C. S., Lee, K. Y., Lee, S. H. (2005) Development of hybrid robot for construction works with pneumatic actuator. Automation in Construction, 14(4):452–459
4. Ostoja-Starzewski, M., Skibniewski, M. (1989) A masterslave manipulator for excavation and construction tasks. Robotics and Autonomous Systems, 4(4):333–337
5. Santos, P.G., Estremera, J., Jimenez, M.A. et al. (2003) Manipulators help out with plaster panels in construction. Industrial Robot, 30(6):508–514
6. Skibniewski, M. J., Wooldridge, S. C. (1992) Robotic materials handling for automated building construction technology. Automation in Construction, 1(3):251–266
7. Masatoshi, H., Yukio, H., Hisashi, M. et al. (1996) Development of interior finishing unit assembly system with robot: WASCOR IV research project report. Automation in Construction, 5(1):31–38
8. Isao, S., Hidetoshi, O., Nobuhiro, T. et al. (1996) Development of automated exterior curtain wall installation system. International Symposium on Automation and Robotics in Construction ISARC'96, Tokyo, Japan
9. Arai, K., N. Miura, E. Muro, E. et al. (1994) Current status and the future direction of construction robotics. 10th annual Organiztion and Management of Building Construction Symposium proceedings. Architectural Institute of Japan
10. Leslie C., Nobuyasu M. (1998) Construction robots; The search for new building technology in japan. ASCE press, Reston
11. Hirabayashi, T., Akizono, J. Yamamoto, T. et al. (2006) Teleoperation of construction machines with haptic information for underwater applications. Automation in Construction, 15(5):563–570
12. Bernold, L.E. (2007) Control schemes for tele-robotic pipe installation. Automation in Construction, 16(4):518–524
13. S.Y. Lee, K.Y. Lee, C.S. Han, et al. (2006) A multi degree-of-freedom manipulator for curtain-wall installation. Journal of Field Robotics, 23(5):347–360
14. S.N. Yu, S.Y. Lee, C.S. Han, et al. (2007) Development of the curtain wall installation robot: Performance and efficiency tests at a construction site. Autonomous Robots, 22(3):281–291

Chapter 2
Control Algorithm for Human–Robot Cooperation

Abstract In this chapter, we suggested a model of the human–robot cooperation system according to the contact conditions, using the adjustable impedance parameters. Also, we structured the whole human–robot cooperative control system by separating into the human impedance control and the experimental impedance control with an inner motion control loop. The experimental contents can be categorized into four areas. We investigated the influences to the system of changes in the impedance parameters, force augmentation ratio, and environmental stiffness parameter. The influence of the inner motion control loop against the system is also described.

Keywords Target dynamics · Impedance parameters · Press fit · Unconstrained case · Constrained case · Force augmentation ratio · Compliance · Stiffness parameter

2.1 Outline

Panels installing operations can be divided into the environment-contacting cases and non-environment-contacting cases. During contact with an environment, it acts as a dynamic constraint and affects an operator. These constraining conditions are usually avoidable through controlling actions, but some of phenomena can be considered as the 'virtual dynamic behaviors' against external forces from the environments including the operator. The mechanical relationship, between an external force and the motion toward the external force, is defined as the impedance, and a desired target is defined as the target dynamics.

From the viewpoint of operational characteristics, the environment-contacting case can be thought as the press fit operation under interactions with panel frames,

S. Lee, *Glazed Panel Construction with Human–Robot Cooperation*,
SpringerBriefs in Computer Science, DOI: 10.1007/978-1-4614-1418-6_2,

which requires relatively higher stability. On the contrary, the non-environment contacting case can be considered as the operation of moving panels promptly to an installation site, which requires relatively higher mobility. In a human–robot cooperative system, we can make an object have impedance characteristics through the use of robot. To get the high stability, the object may have a damping characteristic. However, too much damping decreases the mobility of the system. In this strategy, the impedance parameters of the robot's end-effector are adjusted corresponding to the process of the work of a human operator.

The selection of suitable impedance parameters that guarantee a satisfactory compliant behavior during the interaction may turn out to be inadequate to ensure accurate tracking of the desired position and orientation trajectory when the end effector moves in unconstraint condition. A solution to this drawback can be devised by separating the motion control action from the impedance control action [1]. The motion control action is purposefully made stiffness so as enhance disturbance rejection but, rather than ensuring tracking of a reference position and orientation, it shall ensure tracking of a reference position and orientation resulting from the impedance control action. In other words, the desired position and orientation together with the measured contact force and moment are input to the impedance equation which, via a suitable integration, generates the position and orientation to be used as a reference for the motion control action.

2.2 System Modeling

2.2.1 Operation in Unconstrained Case

The case, in which an operator handles panels on an obstacle-free place, is defined as the operation in unconstrained case. In Fig. 2.1, the force (torque), measured by the operational force sensor which is generated by the interaction between the operator and a panel, is $\underline{\mathbf{F}}_{\mathbf{h}}(\underline{\mathbf{T}}_{\mathbf{h}})$, and the impedance parameters, that are related to a desired dynamic behavior, are $\mathbf{M}_{\mathbf{pt}}(\mathbf{M}_{\mathbf{ot}})$ and $\mathbf{B}_{\mathbf{pt}}(\mathbf{B}_{\mathbf{ot}})$ ($n \times n$ positive definite diagonal inertia and damping matrices) respectively. Here, the desired dynamic behavior of a robot can be given, with the input $\underline{\mathbf{F}}_{\mathbf{h}}(\underline{\mathbf{T}}_{\mathbf{h}})$, by an impedance Eq. 2.1. The subscript 'p' stands for the position and 'o' stands for the orientation, and λ means the force augmentation ratio of an operator.

The **K** (Stiffness Matrices) parameter, having the property of a spring, was excluded as it disturbed the operation to move a panel to a desired position with the operational force. Ultimately, adjusting each of the impedance parameters equals to adjusting the dynamic behavior of a virtual system. The dynamic behavior, generated from the impedance Eq. 2.1 when an operator applies force to a virtual system, is used as a reference that a robot system should follow to move a panel. Table 2.1 shows the inputs and outputs for the modeling of an operator and an environment from the viewpoint of a robot in the unconstrained case. Upon

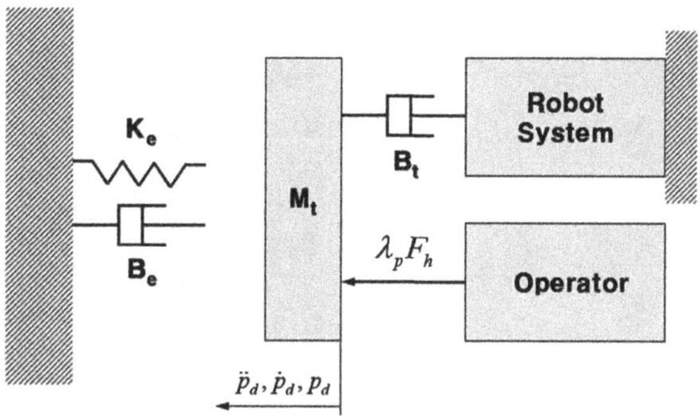

Fig. 2.1 Operation in unconstrained case

Table 2.1 I/O signals in the unconstrained case

	Operator	Environment
Input	$\underline{\mathbf{F}}_{\mathbf{h}}(\underline{\mathbf{T}}_{\mathbf{h}})$	–
Output	$\ddot{\underline{\mathbf{p}}}_{\mathbf{d}}, \dot{\underline{\mathbf{p}}}_{\mathbf{d}}, \underline{\mathbf{p}}_{\mathbf{d}}(\ddot{\underline{\varphi}}_d, \dot{\underline{\varphi}}_d, \underline{\varphi}_d)$	–

input of the operational force, the robot system outputs a desired dynamic behavior.

$$\begin{aligned} \mathbf{M_{pt}}\ddot{\underline{\mathbf{p}}}_{\mathbf{d}} + \mathbf{B_{pt}}\dot{\underline{\mathbf{p}}}_{\mathbf{d}} &= \lambda_p \underline{\mathbf{F}}_{\mathbf{h}} \\ \mathbf{M_{ot}}\ddot{\underline{\varphi}}_d + \mathbf{B_{ot}}\dot{\underline{\varphi}}_d &= \lambda_o \mathbf{T}^{\mathrm{T}}(\varphi_d)\underline{\mathbf{T}}_{\mathbf{h}} \\ \text{where, } \varphi &= [\alpha \quad \beta \quad \gamma]^{\mathrm{T}} \\ \mathbf{T} &= \begin{bmatrix} 0 & -s\alpha & c\alpha s\beta \\ 0 & c\alpha & s\alpha s\beta \\ 1 & 0 & c\beta \end{bmatrix} \end{aligned} \tag{2.1}$$

2.2.2 *Operation in Constrained Case*

In Fig. 2.2, the force (or torque) that was measured by the operational force sensor is $\underline{\mathbf{F}}_{\mathbf{h}}(\underline{\mathbf{T}}_{\mathbf{h}})$, and the force(or torque) that was measured by the experimental force sensor is $\underline{\mathbf{F}}_{\mathbf{e}}(\underline{\mathbf{T}}_{\mathbf{e}})$. With the input values of $\underline{\mathbf{F}}_{\mathbf{h}}(\underline{\mathbf{T}}_{\mathbf{h}})$ and $\underline{\mathbf{F}}_{\mathbf{e}}(\underline{\mathbf{T}}_{\mathbf{e}})$, unlike an operation in the unconstrained case, the desired dynamic behavior of a robot can be described by an impedance Eq. 2.2. For the same reason with case of the unconstrained case, the **K** (Stiffness Matrices), having the property of a spring, was excluded.

$$\begin{aligned} \mathbf{M_{pt}}\ddot{\underline{\mathbf{p}}}_{\mathbf{d}} + \mathbf{B_{pt}}\dot{\underline{\mathbf{p}}}_{\mathbf{d}} &= \lambda_p \underline{\mathbf{F}}_{\mathbf{h}} - \underline{\mathbf{F}}_{\mathbf{e}} \\ \mathbf{M_{ot}}\ddot{\underline{\varphi}}_d + \mathbf{B_{ot}}\dot{\underline{\varphi}}_d &= \mathbf{T}^{\mathrm{T}}(\varphi_d)(\lambda_o \underline{\mathbf{T}}_{\mathbf{h}} - \underline{\mathbf{T}}_{\mathbf{e}}) \end{aligned} \tag{2.2}$$

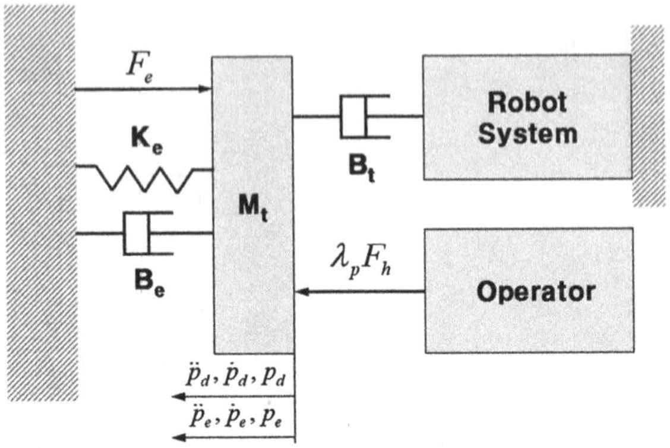

Fig. 2.2 Operation in constrained case

Table 2.2 I/O signals in the unconstrained case

	Operator	Environment
Input	$\underline{\mathbf{F}}_{\mathbf{h}}(\underline{\mathbf{T}}_{\mathbf{h}})$	$\underline{\mathbf{F}}_{\mathbf{e}}(\underline{\mathbf{T}}_{\mathbf{e}})$
Output	$\underline{\ddot{\mathbf{p}}}_{\mathbf{d}}, \underline{\dot{\mathbf{p}}}_{\mathbf{d}}, \underline{\mathbf{p}}_{\mathbf{d}}(\underline{\ddot{\varphi}}_d, \underline{\dot{\varphi}}_d, \underline{\varphi}_d)$	$\Delta\underline{\ddot{\mathbf{p}}}_{\mathbf{de}}, \Delta\underline{\dot{\mathbf{p}}}_{\mathbf{de}}, \Delta\underline{\mathbf{p}}_{\mathbf{de}}$ $(\Delta\underline{\ddot{\varphi}}_{de}, \Delta\underline{\dot{\varphi}}_{de}, \Delta\underline{\varphi}_{de})$

In case interactions with an environment occur, the end effector should endow with a behavior, considering the compliance. In this regard, we defined the relationship between the contact force (or torque) and the position error of the end effector, through the generalized active impedance, as in Eq. 2.3. Thus, the end effector can have linear and dependant impedance characteristics to the translation part, for which the contact force $\underline{\mathbf{F}}_{\mathbf{e}}$ was considered, and the rotation part, for which the equivalent contact moment $\mathbf{T}^{\mathrm{T}}\underline{\mathbf{T}}_{\mathbf{e}}$ was considered. In the Eq. 2.3, $\mathbf{M_{pe}}(\mathbf{M_{oe}})$, $\mathbf{B_{pe}}(\mathbf{B_{oe}})$, $\mathbf{K_{pe}}(\mathbf{K_{oe}})$ are the impedance parameters that determine a dynamic behavior of the end effector for interactions with an environment.

Table 2.2 shows the inputs and outputs for of an operator and an environment from the viewpoint of a robot in the operation in constrained case. Upon input of the operational force and environmental contact force to a robot, the robot system outputs a desired dynamic behavior. The dynamic behavior is determined, as shown in Eq. 2.3, by the environmental contact force and impedance characteristics. The position error indicates the difference between the desired dynamic behavior and the actual dynamic behavior of the end effector. That is to say, it explains that a robot system cannot practically follow a desired dynamic behavior, but indicates a level of compliance with an environment.

$$\begin{aligned}
&\mathbf{M_{pe}}\Delta\underline{\ddot{\mathbf{p}}}_{\mathbf{de}} + \mathbf{B_{pe}}\Delta\underline{\dot{\mathbf{p}}}_{\mathbf{de}} + \mathbf{K_{pe}}\Delta\underline{\mathbf{p}}_{\mathbf{de}} = \underline{\mathbf{F}}_{\mathbf{e}} \\
&\mathbf{M_{oe}}\Delta\underline{\ddot{\varphi}}_{de} + \mathbf{B_{oe}}\Delta\underline{\dot{\varphi}}_{de} + \mathbf{K_{oe}}\Delta\underline{\varphi}_{de} = \mathbf{T}^{\mathrm{T}}(\underline{\varphi}_e)\underline{\mathbf{T}}_{\mathbf{e}} \\
&\text{where, } \Delta\underline{\mathbf{p}}_{\mathbf{de}} = \underline{\mathbf{p}}_{\mathbf{d}} - \underline{\mathbf{p}}_{\mathbf{e}}
\end{aligned} \tag{2.3}$$

2.3 Control Algorithm

As mentioned in Sect. 2.1, the panels installing operation through the human–robot cooperative manipulation can be categorized into the environment-contacting case and the non-environment-contacting case. In the impedance Eq. 2.2, the impedance parameters $\mathbf{M_{pt}}(\mathbf{M_{ot}})$ and $\mathbf{B_{pt}}(\mathbf{B_{ot}})$ are switched to proper values when an operator requires stability or mobility according to the work process. The appropriate parameter values are determined through enough simulations with an experimental system and optimal parameter determination. Also, each of the impedance parameters should be adjusted by stage according to the choice of an operator. This adjusting way is also applied exactly to adjustment of the force augmentation ratio of an operator. The selection of suitable impedance parameters that guarantee a satisfactory compliant behavior during the interaction may turn out to be inadequate to ensure accurate tracking of the desired position and orientation trajectory when the end effector moves without contacting with an environment. A solution to this drawback can be devised by separating the motion control action from the impedance control action.

In order to realize the above solution, it is worth introducing a reference frame other than the desired frame specified by a desired position vector $\underline{\mathbf{p}}_{\mathbf{d}}$ and a desired rotation matrix $\mathbf{R_d}$. This frame is referred to as the compliant frame, and is specified by a position vector $\underline{\mathbf{p}}_{\mathbf{c}}$ and a rotation matrix $\mathbf{R_c}$. In this way, the inverse dynamics motion control strategy can be still adopted as long as the actual end effector position $\underline{\mathbf{p}}_{\mathbf{e}}$ and orientation $\mathbf{R_e}$ is taken to coincide with $\underline{\mathbf{p}}_{\mathbf{c}}$ and $\mathbf{R_c}$ in lieu of $\underline{\mathbf{p}}_{\mathbf{d}}$ and $\mathbf{R_d}$, respectively. Accordingly, the actual end effector linear velocity $\dot{\underline{\mathbf{p}}}_{\mathbf{e}}$ and angular velocity ω_e are taken to coincide with $\dot{\underline{\mathbf{p}}}_{\mathbf{c}}$ and ω_c, respectively. A block diagram of the resulting scheme is sketched in Fig. 2.3 and reveals the presence of an inner motion control loop with respect to the outer impedance control loop. In view Eq. 2.2, the impedance equation is chosen so as to enforce an equivalent mass-damper-spring behavior for the position displacement when the end effector exerts a force (or torque) $\underline{\mathbf{F}}_{\mathbf{e}}$ (or $\underline{\mathbf{T}}_{\mathbf{e}}$) on the environment, i.e.,

$$
\begin{aligned}
&\mathbf{M_{pe}}\Delta\ddot{\underline{\mathbf{p}}}_{\mathbf{dc}} + \mathbf{B_{pe}}\Delta\dot{\underline{\mathbf{p}}}_{\mathbf{dc}} + \mathbf{K_{pe}}\Delta\underline{\mathbf{p}}_{\mathbf{dc}} = \underline{\mathbf{F}}_{\mathbf{e}}\\
&\mathbf{M_{oe}}\Delta\ddot{\underline{\varphi}}_{dc} + \mathbf{B_{oe}}\Delta\dot{\underline{\varphi}}_{dc} + \mathbf{K_{oe}}\Delta\underline{\varphi}_{dc} = \mathbf{T}^{\mathrm{T}}(\varphi_e)\underline{\mathbf{T}}_{\mathbf{e}}\\
&\text{where, } \Delta\underline{\mathbf{p}}_{\mathbf{dc}} = \underline{\mathbf{p}}_{\mathbf{d}} - \underline{\mathbf{p}}_{\mathbf{c}}
\end{aligned}
\tag{2.4}
$$

With reference to the scheme in Fig. 2.3, the impedance control generates the reference position for the inner motion control. Therefore, in order to allow the implementation of the complete control scheme, the acceleration shall be designed to track the position and the velocity of the compliant frame, i.e.,

$$
\begin{aligned}
&\underline{\mathbf{a}}_{\mathbf{p}} = \ddot{\underline{\mathbf{p}}}_{\mathbf{c}} + \mathbf{K_{Dp}}\Delta\dot{\underline{\mathbf{p}}}_{\mathbf{ce}} + \mathbf{K_{Pp}}\Delta\underline{\mathbf{p}}_{\mathbf{ce}}\\
&\underline{\mathbf{a}}_{\mathbf{o}} = \mathbf{T}(\varphi_e)(\ddot{\underline{\varphi}}_{ce} + \mathbf{K_{Do}}\Delta\dot{\underline{\varphi}}_{ce} + \mathbf{K_{Po}}\Delta\underline{\varphi}_{ce}) + \dot{\mathbf{T}}(\varphi_e, \dot{\varphi}_e)\dot{\underline{\varphi}}_{e}\\
&\text{where, } \Delta\underline{\mathbf{p}}_{\mathbf{ce}} = \underline{\mathbf{p}}_{\mathbf{c}} - \underline{\mathbf{p}}_{\mathbf{e}}
\end{aligned}
\tag{2.5}
$$

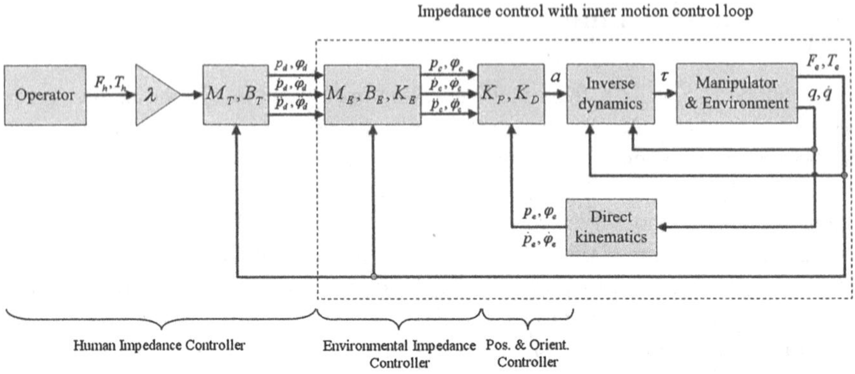

Fig. 2.3 Block diagram of human–robot cooperative manipulation

Notice that $\mathbf{p_c}$ and its associated derivatives can be computed by forward integration of the impedance Eq. 2.4 with input $\underline{\mathbf{F}}_{\mathbf{e}}$ (or $\underline{\mathbf{T}}_{\mathbf{e}}$) available from the multi-axis force/torque sensor.

2.4 Experimental System

As shown in Figs. 2.4 and 2.5, we mounted two sensors in the 2DOF manipulator, moving in the x and y directions. One receives operational signals from an operator, and the other, positioned between the end effector and the 15 kg weighing object, can detect the contact force from the environment. With the signals that are received by the two sensors, the control signals, the manipulator should follow, are generated [2].

Figure 2.6 and Table 2.3 show configuration and components of the control system. The analog output data of the force/torque sensor is sent to an A/D unit and it provides the force (or torque) value to DSP (Dspace Co.ltd). The DSP is used for the force analysis and the impedance control of the manipulator. PC/AT (and display) is used for indicating information containing a real time dynamic behavior of the experimental system. DSP and PC/AT are connected by a local network using LAN cable. DSP sends control information to the PC/AT on real time.

The manipulator is controlled using an impedance control with inner motion loop method based on the force control. It assumes that the manipulator follows a commanded force derived by Eq. 2.6. The sampling time for the force analysis and controlling the manipulator is settled as 1 ms. A mount string was used for the environmental system. The stiffness for an actual environment can be adjusted through replacement of the spring.

$$\begin{bmatrix} \underline{\mathbf{F}}_1 \\ \underline{\mathbf{F}}_2 \end{bmatrix} = \begin{bmatrix} \mathrm{m}_1 + \mathrm{m}_2 & 0 \\ 0 & \mathrm{m}_2 \end{bmatrix} \begin{bmatrix} \ddot{\underline{\mathbf{d}}}_1 \\ \ddot{\underline{\mathbf{d}}}_2 \end{bmatrix} + \begin{bmatrix} \mathrm{m}_1\mathrm{g} + \mathrm{m}_2\mathrm{g} \\ 0 \end{bmatrix} + \begin{bmatrix} \underline{\mathbf{F}}_{\mathbf{ey}} \\ \underline{\mathbf{F}}_{\mathbf{ez}} \end{bmatrix} \tag{2.6}$$

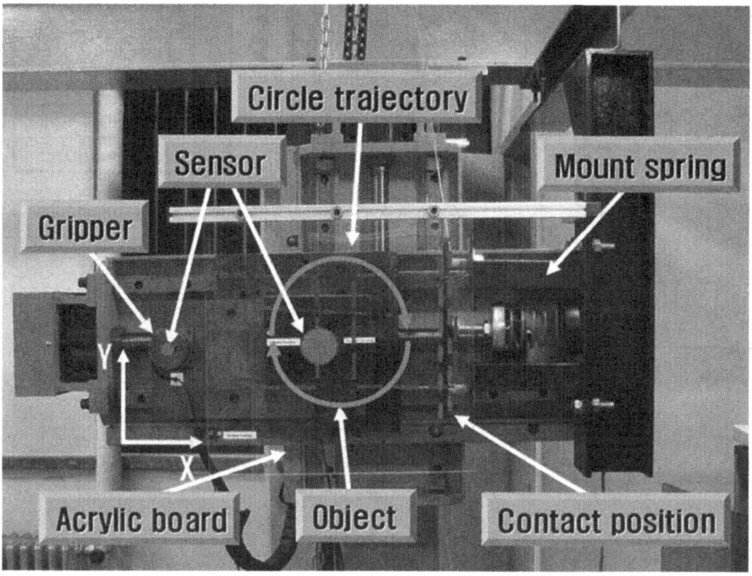

Fig. 2.4 Configuration of an experimental system

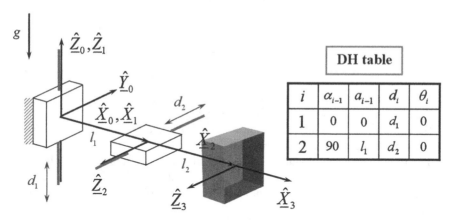

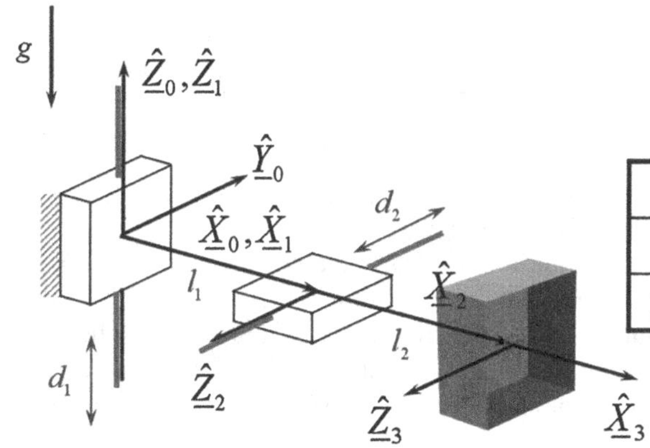

DH table

i	α_{i-1}	a_{i-1}	d_i	θ_i
1	0	0	d_1	0
2	90	l_1	d_2	0

Fig. 2.5 A coordinate system and DH table of the 2DOF manipulator

2.5 Experiment and Results

The experimental methods for the human–robot cooperation-work can be categorized into four staged.

1. An indicator, mounted on an object, automatically moves to the home position from the original position.

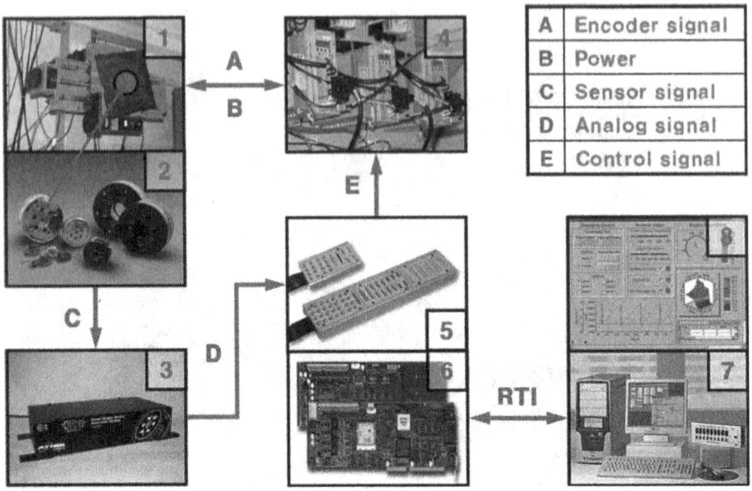

Fig. 2.6 Configuration of a control system

Table 2.3 Components of a control system

No.	Device	No.	Device
1	2DOF manipulator	5	Signal I/O panel
2	Multi-axis F/T sensor	6	Motion controller(DSP)
3	Sensor controller	7	PC/AT
4	Motor controller	8	Graphical user interface

2. The operator applies force to the gripper, so that the indicator follows a circle trajectory that is described on an acrylic board.
3. Based on the operational force, the robot follows the circle trajectory through the impedance control in the unconstraint condition.
4. The robot contacts a mount spring (an environmental system) while following the circle trajectory. The contact force, generated at this time, enables the impedance control, and the robot should endow with a behavior, considering the compliance.

The robot is to follow a circle trajectory, having a diameter of 0.2 m as shown in Fig. 2.7, in almost 180 s. Experiments are conducted after some practices by a healthy managed 29 years. The experimental contents are as follows: Firstly, the influences of each parameter are to be observed for adjustment of the impedance parameters. Secondly, the performance of the suggested impedance control with inner motion control loop is to be evaluated to reduce the position following error for operation of a robot in an unconstrained condition. Thirdly, the influences of $\underline{\mathbf{F}}_{\mathbf{h}}$ and $\underline{\mathbf{F}}_{\mathbf{e}}$, according to change of the force augmentation ratio (λ) of an operator, are to be studied. Finally, the changes of $\underline{\mathbf{F}}_{\mathbf{h}}$ and $\underline{\mathbf{F}}_{\mathbf{e}}$, according to the changes of the actual environmental stiffness, are to be investigated.

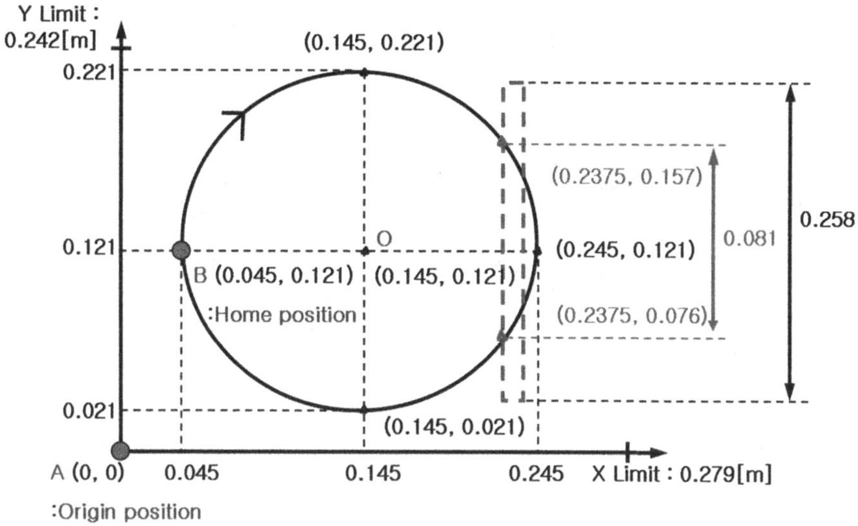

Fig. 2.7 Coordinates of a circle trajectory and each position

2.5.1 Influence of Impedance Parameters

To meet the purpose of this experiment, we performed the experiment through DOE (design of experiment). The factors are given as the impedance parameters $\mathbf{M}_{pt}(\mathbf{M}_{ot})$ and $\mathbf{B}_{pt}(\mathbf{B}_{ot})$ in Eq. 2.2, and the levels are given as 1, 3, and 10. The characteristic values according to results of the experiment are described in Figs. 2.8 and 2.9. The force, used for the two experiments, is 5 N and applied for around 10 s. The $\mathbf{K}$ value is to be set to 1 N/m for convergence of graphs. Figure 2.8 shows a case, where the factor is $\mathbf{M}_{pt}$, which is related to an operation that requires mobility (i.e. response velocity of a robot) in panels-handling operational process. That is to say, this operation does not require relatively higher stability, but requires prompt response velocity of a robot with small operational force. As the $\mathbf{M}_{pt}$ value rises, a robot has low response velocity with respect to same operational force.

Figure 2.9 shows a case, where the factor is $\mathbf{B}_{pt}$, which is related to an operation that requires stability (i.e., anti-malfunction due to high velocity) in panels-handling operational process. That is to say, this case does not require relatively higher mobility, but requires a precise and stable operation. As $\mathbf{B}_{pt}(\mathbf{B}_{ot})$ value rises, the robot's motion velocity is decreased even the same operational force. As the stability is increased, the more demanding force may make an operator feel the minimum moving distance shorter.

The two factors are related to mobility and stability, and interact with each other. Therefore, these two factors should be traded off appropriately so that a robot system can have the maximum mobility in the range of securing the system

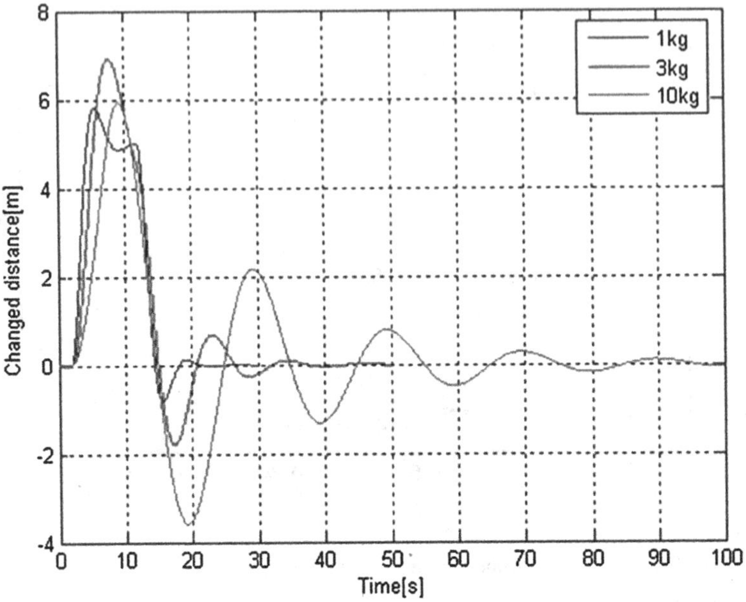

Fig. 2.8 Influence of impedance parameter $M_{pt}(M_{ot})$

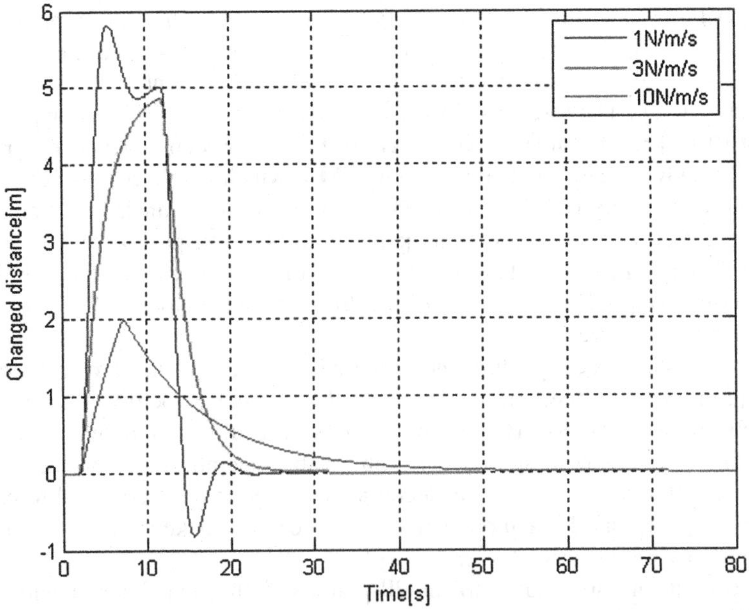

Fig. 2.9 Influence of impedance parameter $B_{pt}(B_{ot})$

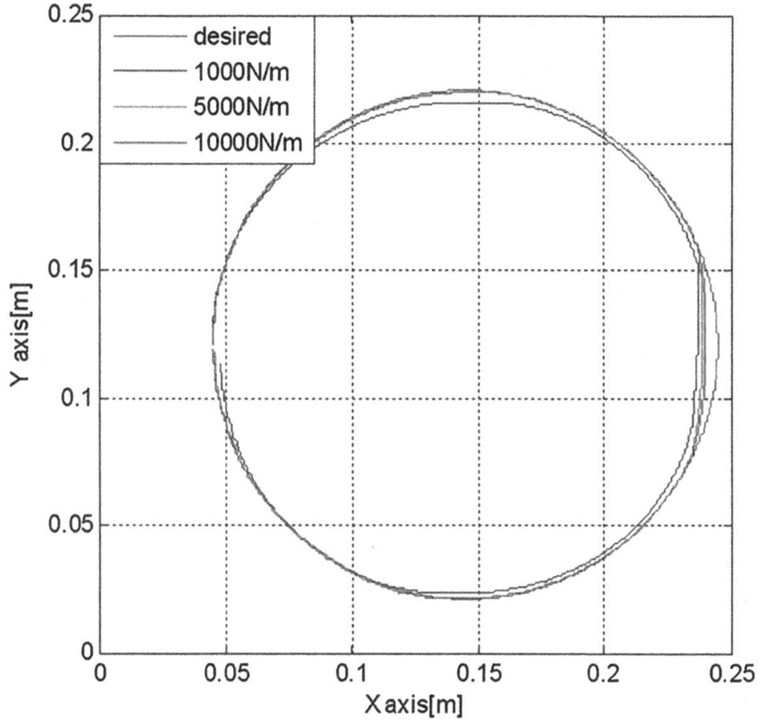

Fig. 2.10 Motion trajectories without inner motion control loop

stability or vice versa. Also, stage-by-stage adjustment of each factor should be available in consideration of an operator's choice. The $\mathbf{M_{pt}}$ and $\mathbf{B_{pt}}$ values, used in the experiments, were 50**I** and 2500**I** respectively for the mobility-requiring operation (unconstraint case), and 15**I** and 7500**I** respectively for the stability-requiring operation (constraint case).

2.5.2 Influence of Inner Motion Control Loop

Figure 2.10 shows a graph that was achieved by applying changes of the environmental stiffness factor ($\mathbf{K_{pe}}$) to the proposed impedance control system. Without the operational force, the constant force, which is input to the controller, makes the robot follow an already-programmed circle trajectory (desired). As shown in the graph, it can be recognized that the path tracking accuracy is rather poor during execution of the whole tasks if the stiffness parameter is small. The small stiffness parameter also causes reduction of the contact force in the constraint condition. These results occur due to a larger end effector position error in operation.

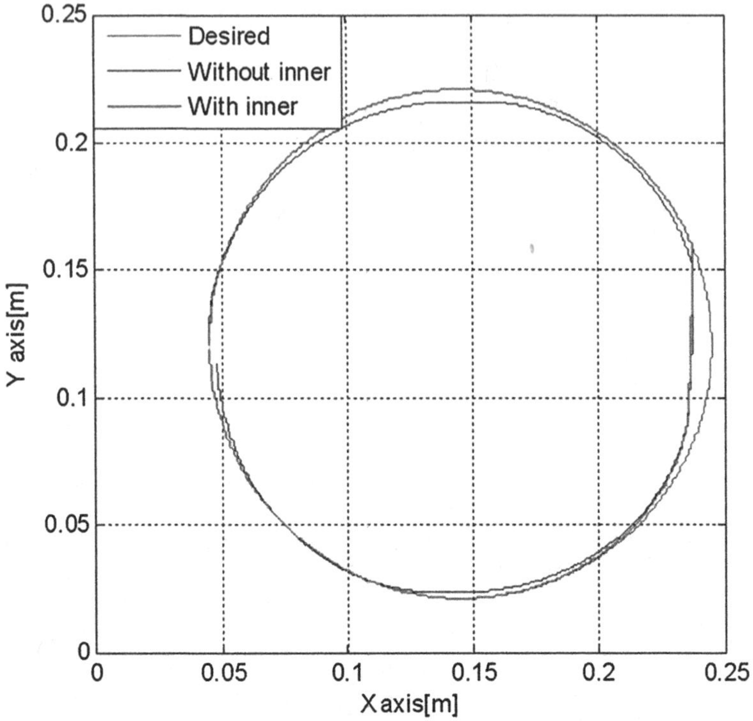

Fig. 2.11 Motion trajectories with inner motion control loop

To solve these problems, we proposed the impedance control with an inner motion control loop. The inner motion loop gains in Eq. 2.5 have been set as $\mathbf{K_{Dp}} = 1.5\mathbf{I}$ and $\mathbf{K_{Pp}} = 15\mathbf{I}$. Figure 2.11 shows a graph that compares a case with the inner motion control loop and a case without the inner motion control loop when the environmental stiffness parameter is 1,000 N/m. As indicated in the graph, the robot operation without the inner motion control loop shows inferior desired-position-following performance in an unconstraint condition. By using the inner motion control loop, the high following performance can be obtained as shown in the Fig. 2.11.

2.5.3 Influence of Force Augmentation Ratio

Figure 2.12 shows a result graph that was obtained from the suggested experimental method for human–robot cooperation-work. The operator applies force to the gripper so that the indicator, mounted in the object, can follow the circle trajectory that is described on the acrylic board. In the graph, the part A is the area where the object contacts the environment (mount spring). The compliance,

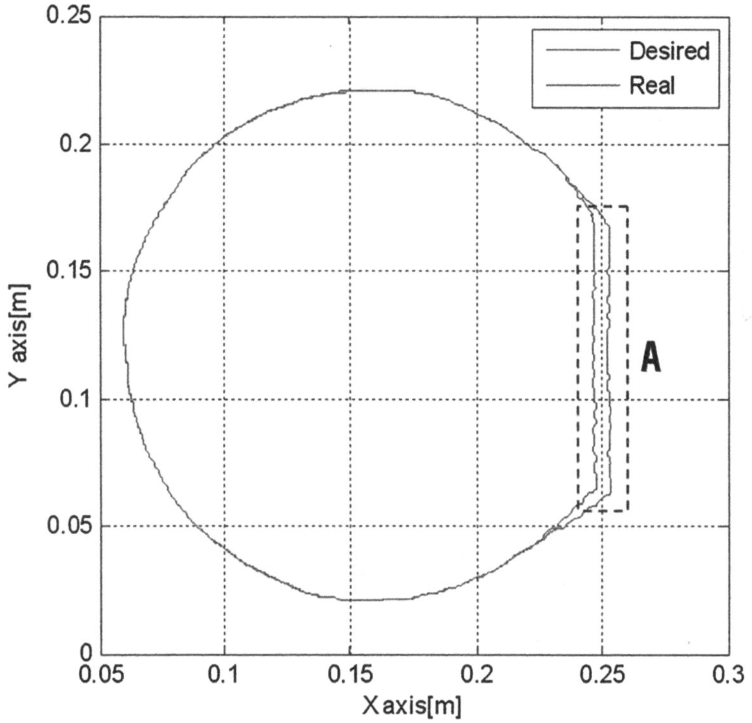

Fig. 2.12 Robot trajectory with compliant motion

determined by the impedance parameters of the environment, is provided, and the glazed panels handling operation is completed while the robot system and the environment are not damaged.

The force augmentation ratio (λ), suggested in Eq. 2.1 and Eq. 2.2, plays a role of controlling the scale of the force that is required by an operator for a human–robot cooperation-work. In the experiments, we studied changes of $\underline{\mathbf{F}}_{\mathbf{h}}$ and $\underline{\mathbf{F}}_{\mathbf{e}}$ when the force augmentation ratio (λ) was increased from 3 to 6. The impedance parameters for the experiments were set as $\mathbf{M}_{\mathbf{pt}} = 15\mathbf{I}$, $\mathbf{B}_{\mathbf{pt}} = 7500\mathbf{I}$, $\mathbf{M}_{\mathbf{pe}} = 50\mathbf{I}$, $\mathbf{B}_{\mathbf{pe}} = 10000\mathbf{I}$, and $\mathbf{K}_{\mathbf{pe}} = 8000\mathbf{I}$. Figures 2.13 and 2.14 show the graphs of $\underline{\mathbf{F}}_{\mathbf{h}}$ and $\underline{\mathbf{F}}_{\mathbf{e}}$, used to obtain the graph in Fig. 2.12.

As shown in Fig. 2.13 ($\lambda = 3$), $\underline{\mathbf{F}}_{\mathbf{h}}$, for handling the object, is required to be around 10 N in case of no contact with the environment, while $\underline{\mathbf{F}}_{\mathbf{h}}$ is required to be 30 N in case of contact with the environment. As λ increased to 6 from 3, $\underline{\mathbf{F}}_{\mathbf{h}}$, required in case of no contact with the environment, was reduced by a half, and $\underline{\mathbf{F}}_{\mathbf{h}}$ for the environment-contacting case was also reduced to 15 N by around a half. We can see that the force, required by an operator, gets smaller as λ increases, but there is no significant change in the force ($\underline{\mathbf{F}}_{\mathbf{e}}$) that reflects in the contacting condition.

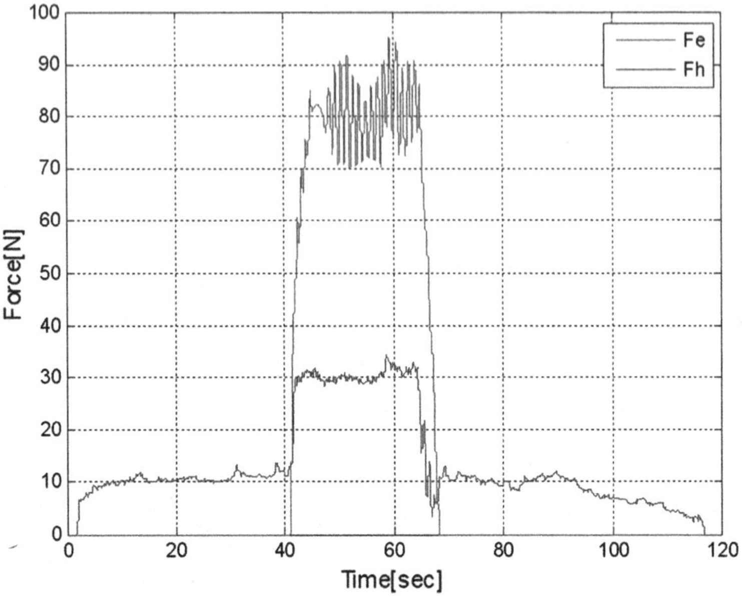

Fig. 2.13 $\underline{F}_h$ and $\underline{F}_e$ at $K_{pe} = 8000I$ and $\lambda = 3$

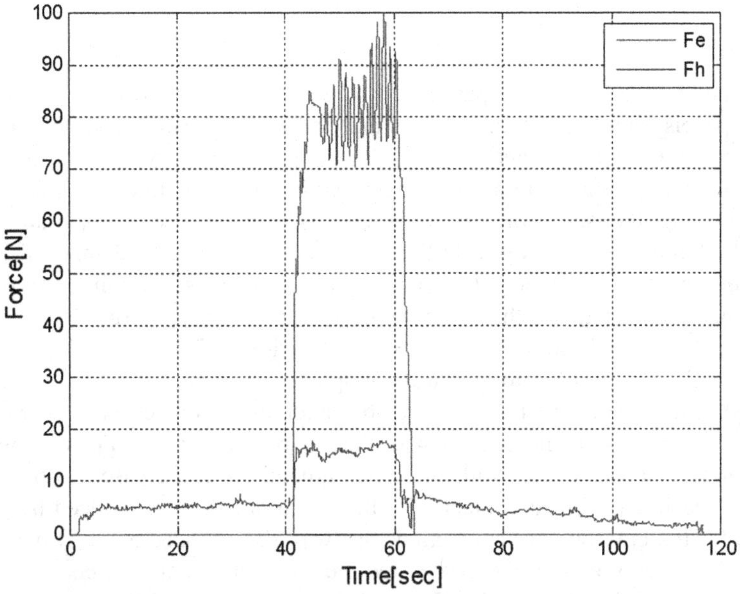

Fig. 2.14 $\underline{F}_h$ and $\underline{F}_e$ at $K_{pe} = 8000I$ and $\lambda = 6$

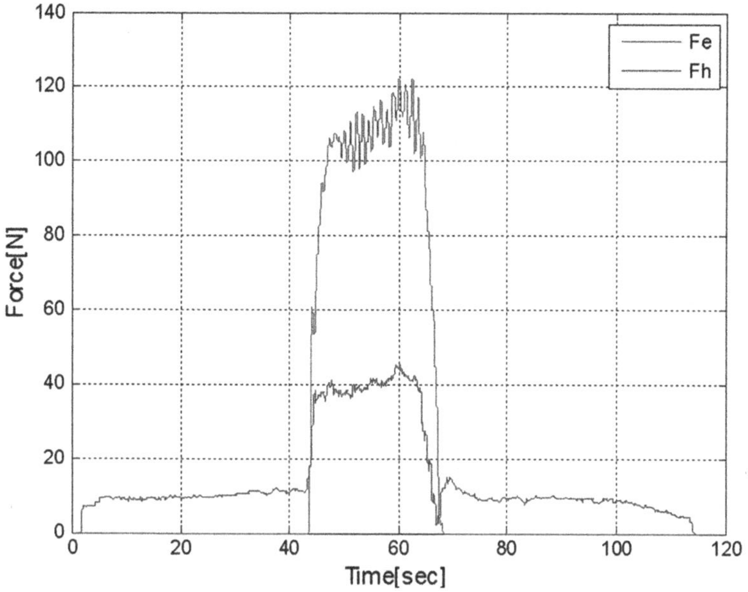

Fig. 2.15 F_h and F_e at $K_{pe} = 11000I$ and $\lambda = 3$

2.5.4 Influence of an Environmental Stiffness Parameter

The environmental stiffness ($\mathbf{K_e}$) depends on the characteristics of materials, composing an environment. The purpose of this experiment, changing the environmental stiffness, lies on comparison of the reaction force ($\underline{\mathbf{F}}_\mathbf{e}$) that is felt by an operator according to operational conditions such as a case contact with an obstacle occurs or a case the press fit is required. The impedance parameters for the experiment were set as $\mathbf{M_{pt}} = 15\mathbf{I}$, $\mathbf{B_{pt}} = 7500\mathbf{I}$, $\mathbf{M_{pe}} = 50\mathbf{I}$, $\mathbf{B_{pe}} = 10000\mathrm{I}$, and, $\mathbf{K_{pe}} = 8000\mathbf{I}$. Figures 2.15 and 2.16 show the graphs of $\underline{\mathbf{F}}_\mathbf{h}$ and $\underline{\mathbf{F}}_\mathbf{e}$, used to obtain the graph in Fig. 2.12.

As shown in Fig. 2.15 ($\mathbf{K_{pe}} = 11000\mathbf{I}$), $\underline{\mathbf{F}}_\mathbf{h}$ of around 10 N is required to move the object in case there is no contact with the environment. In case there is contact, however, $\underline{\mathbf{F}}_\mathbf{h}$ (required) is around 40 N and $\underline{\mathbf{F}}_\mathbf{e}$ (generated) is around 100 N. We could find that there was no change in $\underline{\mathbf{F}}_\mathbf{h}$, for handling the object, in case of no contact with the environment, while $\underline{\mathbf{F}}_\mathbf{h}$ and $\underline{\mathbf{F}}_\mathbf{e}$, for moving the object, increased from 30 N to 40 N and from 80 N to 110 N respectively in case of contact with the environment, when $\mathbf{K_{pe}} = 8000\mathbf{I}$.

A similar result is shown when λ increases from 3 to 6 (Fig. 2.16). In case there is no contact with the environment, $\underline{\mathbf{F}}_\mathbf{h}$ of around 5 N is required to move the object. In case there is contact, however, the required force $\underline{\mathbf{F}}_\mathbf{h}$ is around 20 N and the generated force $\underline{\mathbf{F}}_\mathbf{e}$ is around 110 N. We could find that there was no change in $\underline{\mathbf{F}}_\mathbf{h}$, for handling the object, in case of no contact with the

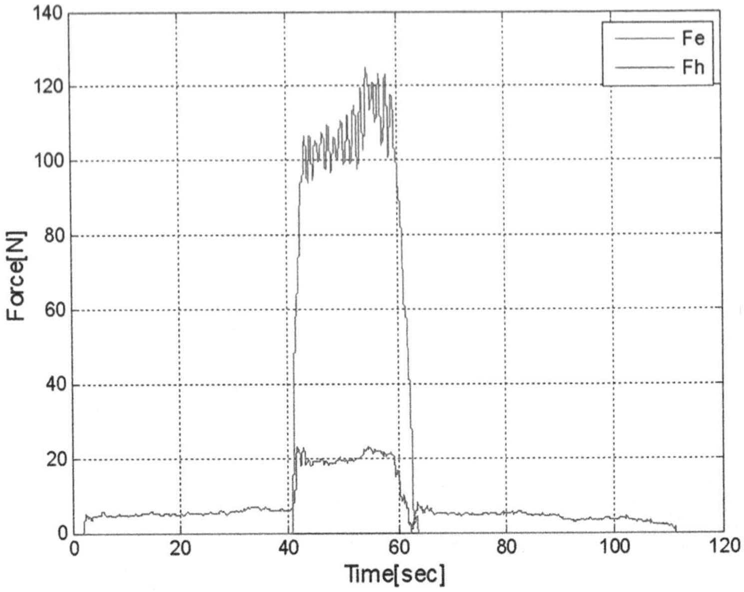

Fig. 2.16 F_h and F_e at $K_{pe} = 11000I$ and $\lambda = 6$

environment, while $\underline{\mathbf{F}}_{\mathbf{h}}$ and $\underline{\mathbf{F}}_{\mathbf{e}}$, for moving the object, increased from 15 N to 20 N and from 80 N to 110 N respectively in case of contact with the environment, when $\mathbf{K_{pe}} = 8000\mathbf{I}$. It can be recognized that, as $\mathbf{K_{pe}}$ increases, the force, required by an operator, gets increased and the force ($\underline{\mathbf{F}}_{\mathbf{e}}$), reflecting in the contact condition, also gets increased.

References

1. Bruno S., Luigi V. (1999) Robot force control. Kluwer Academic Publishers, Boston
2. S.Y. Lee, K.Y. Lee, S.H. Lee, et al. (2007) Human-robot cooperation control for installing heavy construction materials. Autonomous Robots, 22(3):305–319

Chapter 3
Conceptual Design of Human–Robot Cooperative System

Abstract A discussion follows that describes the conceptual design of a human–robot cooperative system, which is proposed to construct glazed ceiling panels. First, considering the environment where the system is used, the required specification about the target building that is to be heavy glazed ceiling panels should be constructed. Then, the existing construction equipment and construction process for installing heavy glazed ceiling panels are analyzed. According to analysis of the target work and building, the functional requirements for implementing a glazed ceiling panel construction robot are deduced. Last, a conceptual design was produced for the proposed robot.

Keywords Glazed ceiling panels · Functional requirements · Conceptual design · Falling accidents · Vehicle rollovers · Musculo–skeletal disorders · Module T&H-bar · Aerial work platform · Multi-DOF manipulator · Human–robot interface · Vacuum pads

3.1 Jab Analysis

Building materials and components are much larger and heavier than many other industrial materials. Buildings are made of many kinds of materials and each material may be a different shape. A glazed ceiling panel is one type of building material for interior finishing. The demand for larger glazed ceiling panel has been increasing along with the number of high-rise buildings and the increased interest in interior design. To introduce robotic technology for installing glazed ceiling panel on construction site, we should be considered some problems. Glazed ceiling panel construction robots are receiving special attention because of the difficulties of transporting the panel to high installation positions and handling the fragile

S. Lee, *Glazed Panel Construction with Human–Robot Cooperation*,
SpringerBriefs in Computer Science, DOI: 10.1007/978-1-4614-1418-6_3,

Fig. 3.1 Ceiling panel construction process

building material. In order to address these conditions, the form of a glazed ceiling panel construction robot is different from other construction robots.

The existing ceiling panel construction process, which is complicated and hazardous, relies on scaffolding (or aerial lift) and human labor as shown in Fig. 3.1. This process exposes operators to falling accidents or vehicle rollovers. In addition, inappropriate working posture is a major element that increases the frequency of accidents by causing various musculo–skeletal disorders and decreasing concentration. That is to say, it becomes a direct cause of decreasing productivity and safety in construction.

Before the conceptual design, considering the environment where a construction robot is used, the required specification about the target building that is to be heavy glazed ceiling panels should be constructed. Figure 3.2 shows the target construction site and panel installation position (soffit of building) related to this study. The building size is 32 × 22 m and the installation position of the glazed ceiling panel is 15 m above the ground. The glazed ceiling panel measures 3,000 × 1,500 mm (the maximum size) and weighs max. 160 kg. The construction method is 'Module T&H-bar', which represents the 'Lay-in' to place the glazed ceiling panel on ceiling frames.

Figures 3.2 and 3.3 show the existing construction equipment and construction process for installing heavy glazed ceiling panels. This process exposes operators to falling accidents or vehicle rollovers. In addition, inappropriate working posture is a major element that increases the frequency of accidents by causing various musculo–skeletal disorders and decreasing concentration. That is to say, it becomes a direct cause of decreasing productivity and safety in construction (Fig. 3.4).

3.2 System Configuration

According to analysis of the target work and building from the previous section, it is deduced that the functional requirements for implementing a glazed ceiling panel construction robot are as follows. It can be divided into the two aspects

Fig. 3.2 Soffit: the underside of a projecting floor (*dotted line*)

Fig. 3.3 Existing construction equipment

of approaching the solution: hardware and software. In the side of hardware, the solution includes the aerial work platform to reach the height of the workplace, the multi-Degree of Freedom (DOF) manipulator to replace the workforce, and lastly the human–robot interfacing device to make a human cooperate with a robot interactively which includes the end-effector to handling the slippery surface of glazed panels. In the side of software, the human–robot cooperative system which follows the command from the skillful worker and assists the labor with the amplified robotic force and complies with the external environment like a contact force is applied to the entire control algorithm.

1. The glazed ceiling panel construction robot must be able to lift heavy glazed ceiling panels, an operator, and the installation equipment. It requires engines, batteries, or motors to lift the weight.

Fig. 3.4 Existing construction process. **a** Loading, **b** Moving, **c** Installing, **d** Finishing

2. The glazed ceiling panel construction robot must be able to handle heavy and fragile glazed ceiling panels. It requires sophisticated force and position control including human–robot cooperative control. That is to say, the operator must be able to perceive external information that is received by the robot.
3. The glass ceiling installation robot must be devised to help operators, not to replace them. This requires a smart Human–Robot Interface (HRI) to interact with operators and robots. The robot must share the work space with an operator.
4. The glass ceiling installation robot must be able to reflect the technical operator's skills that are required to obtain homogeneous construction quality. Thus, the robot must follow the operator's intentions in various environments at unstructured construction sites.
5. The glass ceiling installation robot must belong in the working process. This is required to prevent operator accidents and help operators increase productivity, by reducing the recovery time from accidents and increasing the operator's duty time.

According to analysis of the upper functional requirements, a conceptual design was generated as below, which will influence the suggested robot (i.e., glazed ceiling panel construction robot).

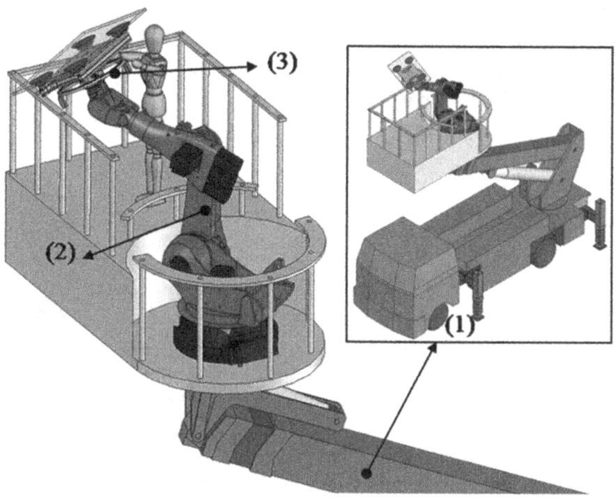

Fig. 3.5 Conceptual diagram of the glazed ceiling panel construction robot. (1) Truck mounted type aerial work platform, (2) Six-DOF manipulator, (3) Human–robot interfacing device including vacuum pads

1. An aerial work platform is needed that can support heavy glazed ceiling panels, an operator, and the installation equipment with enough work space to reach about 15 m above the ground.
2. A multi-DOF manipulator is required to install heavy glazed ceiling panels, which replaces a large amount of human labor. The manipulator has to be chosen according to work space and payload.
3. This robot is a semi-automated system to cope with a constantly changing work environment. Thus, the robot works and coexists with operators in atypical work conditions. The operator uses the multi-DOF manipulator from the deck of an aerial lift.
4. The gripper is based on vacuum and mechanical devices, while the glazed ceiling panel gripping is performed manually.
5. After gripping the glazed ceiling panel, the manipulator is operated by an operator. An operator supplies external force containing an operational command on the HRI device. Therefore, the manipulator can be controlled by an intuitive installation method that can reflect the dexterity of a technical operator.
6. The deck of an aerial work platform serves simultaneously as an operator's work space and a manipulator. Therefore, the operator's safety and productivity is influenced by the design of the deck.
7. The control strategy of the suggested robot is a combination of the force applied and the robot.

From the above approach, the conceptual diagram can be drawn as Figs. 3.5 and 3.6. The details advances to realize the construction robot by selecting the appropriate multi-DOF manipulator, applying the human–robot interface system (HRIs) based on the advanced research, adopting the aerial work platform with modifications for our application, and making the control system to cover the entire system. The important thing that should be considered to select the

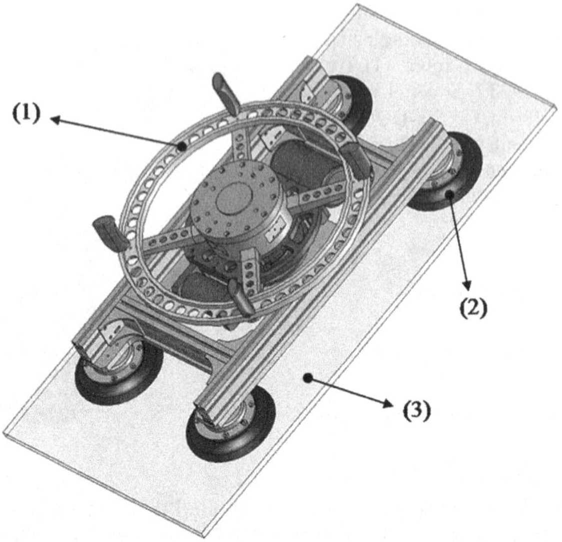

Fig. 3.6 Human–robot interfacing device. **a** Human–robot interfacing handle, **b** Vacuum pans, **c** Glazed ceiling panel

Table 3.1 The required specification for glazed ceiling panel construction

Specification	Descriptions
Aerial work platform	Height of the workplace of 15 m
Allowable weight (Payload)	A labor: approx. 70 kg
	A ceiling glass: approx. 480 kg (160 kg × 3EA)
	Equipment: dependent on the design
Construction method	Lay-in method installing the glazed ceiling panel (the existing construction method)
Robotic gripper	Vacuum pad type handler
Robot operational method; Human–robot cooperation	Intuitive manipulation
	Power assistance
	Force reflection

appropriate multi-DOF manipulator is payload. The panel's weight to be our target object is 80 $\sim$ 160 kg. Taking the HRIs into account, the required payload is approximately 200 kg. In this application, the multi-DOF manipulator from KUKA Roboter GmbH, KR200 comp model is adopted as shown in Fig. 3.5.

The human–robot interface system connects a labor to robotic system physically. It transfers a human's command signals from force/torque sensor to the robotic system to follow the motions generated intuitively by operator. Also, it receives the signals from the environment through another force/torque sensor and processes to make an operator feel force reflected from the robot's end-effector acting with working environment. As shown in Fig. 3.6, the interfacing device includes the human–robot cooperative system and the vacuum pads to grip the glazed ceiling panel.

In aspect of control strategy, the modeling of human–robot relation is derived from the target dynamic system with impedances and the robot–environment relation is modeled by the impedance system. The impedance controller with position control is chosen to improve the traceability while the robot works on the unconstrained case (no contact between the robot and the environment).

After generating the conceptual design, the specifications to satisfy the construction robot to be developed are derived as shown in Table 3.1. First of all, the customized construction equipment is necessary to replace the existing constructing manpower to handle and install the heavy weighed panels safely. Also, the construction method with robotic process should reflect the same quality as the existing method does which ensure the stabilized construction quality. In addition, to improve the productivity, the measures based on the process design of optimized installing procedure, the simulation, and the mock-up test should be taken prior to the application in real field.

Chapter 4
Prototype for Glazed Panel Construction Robot

Abstract The prototype system of human–robot cooperative manipulation presented in this chapter combines a mobile platform and a manipulator standardized in modular form to compose its basic system. Also, the hardware and software necessary for each area of application were composed of additional modules and combined with the robot's basic system. The suggested prototype can execute particular operations in various areas such as construction, national defense and rescue by changing these additional modules.

Keywords Mobile platform · Denavit-hartenberg notation · Lagrange formulation · Virtual-work theory · Force/torque sensor · Wired/wireless control · Outrigger

4.1 Basic System

A series-type 6DOF manipulator and a 3DOF mobile platform were suggested for use in the basic system of prototype, in consideration of the workspace and mobility. However, it is possible to change the elements of the basic system according to load specifications.

4.1.1 6DOF Manipulator

Figure 4.1 shows the 6DOF manipulator (Samsung Electronics Co.ltd). This robot is a special case manipulator where the centers of the last three axes meet in the center of the robot wrist. The kinematic analysis in such form of manipulator can

S. Lee, *Glazed Panel Construction with Human–Robot Cooperation*,
SpringerBriefs in Computer Science, DOI: 10.1007/978-1-4614-1418-6_4,

Fig. 4.1 6DOF manipulator

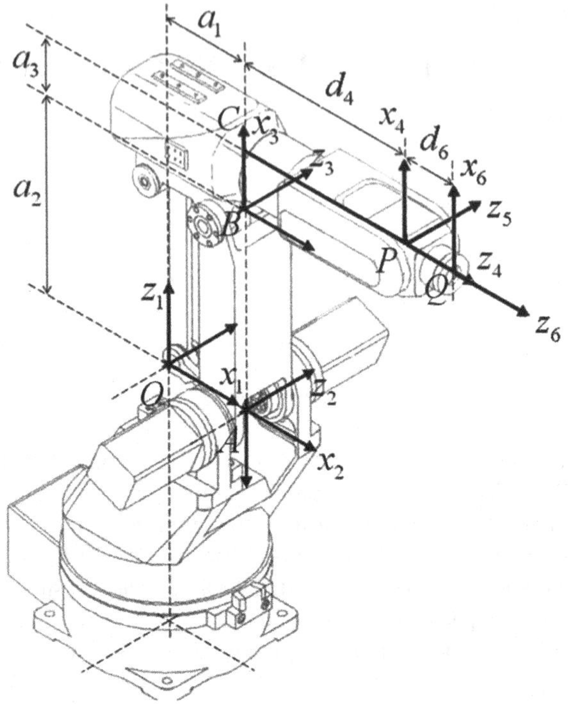

Table 4.1 Specifications of 6DOF manipulator

Specification	Value
Degree of freedom	6
Weight capacity	58.38 N
Arm length(max)	858 mm
Velocity of end-effector	30°/s
Weight	588 N

be divided into two link chains (the first three link chains and then the other three link chains). Table 4.1 shows the specifications of the manipulator.

The forward kinematics of the manipulator is defined by the question of solving for the position and direction of the end-effector according to each degree of the joints. That is, it is the problem of solving the position vector and rotational matrix of the end-effector. The kinematic analysis of the manipulator can be executed with any coordinate system but the most typical one is Denavit-Hartenberg Notation (noted as D-H notation below). The unknown kinematics values of a series manipulator can be solved by multiplying the homogeneous matrix defined in the following Eq. 4.1 and solved similar to Eq. 4.2.

Table 4.2 D-H parameters of 6DOF manipulator

i	$\alpha_i(°)$	a_i(mm)	d_i(mm)	$\theta_i(°)$
1	−90	a_1	0	θ_1
2	0	a_2	0	θ_2
3	−90	a_3	0	θ_3
4	90	0	d_4	θ_4
5	−90	0	0	θ_5
6	0	0	d_6	θ_6

$$\mathbf{H} = \mathbf{R}(\theta_i, z)\mathbf{T}(d_i, z)\mathbf{T}(a_i, x)\mathbf{R}(\alpha_i, x)$$
$$= \begin{bmatrix} \cos\theta_i & -\sin\theta_i\cos\alpha_i & \sin\theta_i\sin\alpha_i & a_i\cos\alpha_i \\ \cos\theta_i & \cos\theta_i\cos\alpha_i & -\cos\theta_i\sin\alpha_i & a_i\sin\alpha_i \\ 0 & \sin\alpha_i & \cos\alpha_i & d_i \\ 0 & 0 & 0 & 1 \end{bmatrix} \tag{4.1}$$

$${}^{n}_{0}\mathbf{H} = {}^{1}_{0}\mathbf{H}\cdot{}^{2}_{1}\mathbf{H}_i\cdots{}^{n}_{n-1}\mathbf{H}_i = \begin{bmatrix} {}^{n}_{0}\mathbf{R} & {}^{0}\underline{\mathbf{P}}_n \\ 0 & 1 \end{bmatrix}$$

$${}^{0}_{6}\mathbf{H} = \begin{bmatrix} u_x & v_x & w_x & q_x \\ u_y & v_y & w_y & q_y \\ u_z & v_z & w_z & q_z \\ 0 & 0 & 0 & 1 \end{bmatrix} \tag{4.2}$$

Table 4.2 shows the D-H parameters through the D-H notation using the coordinate system defined in Fig. 4.1. The position of the end-effector can be calculated by substituting a given parameter into the D-H transformation matrix in Eq. 4.2.

Analysis of the inverse kinematics of this manipulator is done by separating the information of the first link chain from that of the second link chain using the center position of the wrist (P). Equation 4.3 shows the position from the coordinate system of the end-effector to the center of the wrist and Eq. 4.4 shows the position seen from the initial coordinate system. In addition, Eq. 4.5 shows the position seen from the center of the wrist.

$${}^{6}\underline{\mathbf{p}} = \overline{\mathrm{QP}} = [0,\ 0,\ -\mathrm{d}_6,\ 1]^{\mathrm{T}} \tag{4.3}$$

$${}^{0}\underline{\mathbf{p}} = \overline{\mathrm{OP}} = \begin{bmatrix} \mathrm{p_x} \\ \mathrm{p_y} \\ \mathrm{p_z} \\ 1 \end{bmatrix} = \begin{bmatrix} q_x - \mathrm{d}_6 w_x \\ q_y - \mathrm{d}_6 w_y \\ q_z - \mathrm{d}_6 w_z \\ 1 \end{bmatrix} \tag{4.4}$$

$${}^{3}\underline{\mathbf{p}} = \overline{\mathrm{CP}} = [0,\ 0,\ \mathrm{d}_4,\ 1]^{\mathrm{T}} \tag{4.5}$$

Equations 4.4 and 4.5 can be arranged as

$$ {}^{0}\underline{\mathbf{p}} = {}^{0}_{3}\mathbf{T}\,{}^{3}\underline{\mathbf{p}} \tag{4.6} $$

The information on the first link chain was decoupled as Eq. 4.7 using Eq. 4.6.

$$ \left({}^{0}_{1}\mathbf{R}\right)^{-1}{}^{0}\underline{\mathbf{p}} = {}^{1}_{3}\mathbf{R}\,{}^{3}\underline{\mathbf{p}} \tag{4.7} $$

Forward and inverse kinematics can be analyzed by arranging the left and right sides of Eq. 4.7.

Generally, dynamic modeling of a manipulator can be derived like Eq. 4.8 through a Lagrange formulation.

$$ \mathbf{B}(q)\ddot{\mathbf{q}} + \mathbf{C}(q,\dot{q})\dot{\mathbf{q}} + \mathbf{g}(q) = \underline{\xi} \tag{4.8} $$

Each joint torque of a manipulator is produced by a direct drive or gear drive, and consists of a form similar to Eq. 4.9.

$$ \underline{\xi}_i = \underline{\tau}_i - \underline{\tau}_{fi} - \underline{\tau}_{ei} \tag{4.9} $$

$\underline{\tau}_{fi}$ is the torque derived from joints and shows the torque produced by friction from the joints. Also, $\underline{\tau}_{ei}$ shows the power and moment from the outside affected when the end-effector contacts the surrounding environment. Friction from the joints is a simplified model and only viscous friction is considered. That is,

$$ \underline{\tau}_f = \mathbf{F}\dot{\underline{\mathbf{q}}} \tag{4.10} $$

Here, $\mathbf{F}$ shows the viscous friction coefficient from each joint.

Lastly, if $\underline{\mathbf{f}}$ is defined as the external force vector applied to the end-effector and $\underline{\mu}$ is defined as the external torque vector applied to the end-effector, the following joint torque value can be calculated by applying the 'virtual-work theory'.

$$ \underline{\tau}_e = \mathbf{J}^{\mathrm{T}}(q)\underline{\mathbf{h}} \quad \text{where, } \underline{\mathbf{h}} = \begin{bmatrix} \underline{\mathbf{f}} \\ \underline{\mu} \end{bmatrix} \tag{4.11} $$

From Eq. 4.11, therefore, the final suggested dynamic modeling of a 6DOF manipulator can be shown as Eq. 4.12.

$$ \mathbf{B}(q)\ddot{\underline{\mathbf{q}}} + \mathbf{C}(q,\dot{q})\dot{\underline{\mathbf{q}}} + \mathbf{F}\dot{\underline{\mathbf{q}}} + \mathbf{g}(q) = \underline{\tau} - \mathbf{J}^{\mathrm{T}}(q)\underline{\mathbf{h}} \tag{4.12} $$

4.1.2 Mobile Platform

A 6DOF manipulator is fitted to the top plate of the mobile platform. Thus, movement of the manipulator is possible according to the platform's DOF. Also,

Table 4.3 Specifications of mobile platform

Specification	Value
Maximum load of carriage	9,800 N
Weight	3,920 N
Length	1,260 mm
Breadth	900 mm
Velocity	Maximum 2.5 km/h
Inclination of degree	20°
Power consumption	0.8 kW
Source of electricity	Charging battery

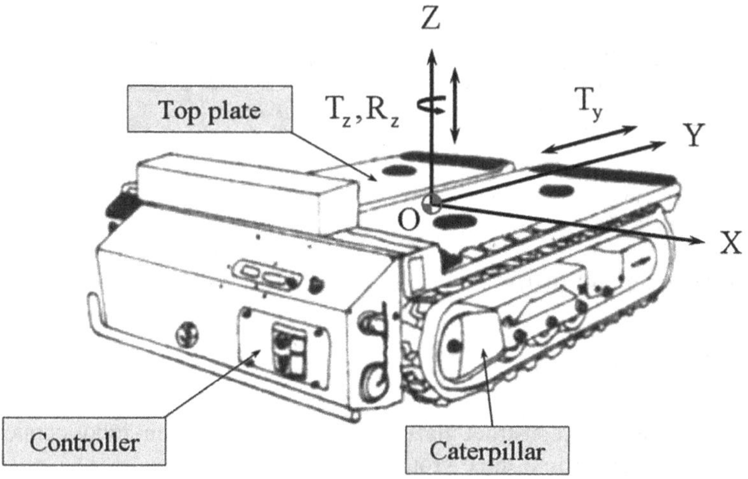

Fig. 4.2 Mobile platform (caterpillar type)

traveling on uneven surfaces or surfaces with barriers is made possible using caterpillar tread. Table 4.3 shows the specifications of the suggested mobile platform (Kajima Mechatro Engineering Co.ltd). This mobile platform largely consists of caterpillar tread, a top plate and a controller as shown in Fig. 4.2. The caterpillar tread is powered by two DC motors with a reduction gear. Also, perpendicular movement of the top plate is achieved through a hydraulic cylinder. Through such movement mechanisms, 3DOF movement of forward and backward (T_Y), left and right rotation (R_Z) and perpendicular movement of the top plate is realized with the central axis (Z) of the mobile platform as the base. It is possible to control movement through both wired and wireless controllers and traveling speed can be controlled through an internal controller. The details of specification are shown in Table 4.3.

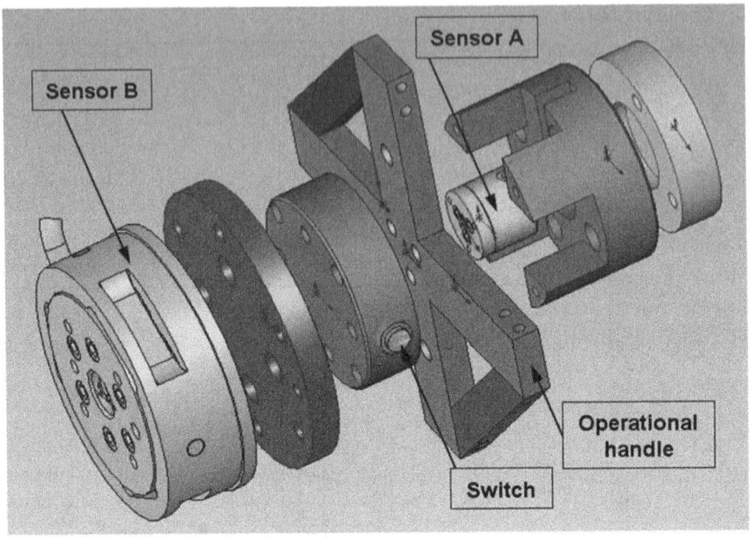

Fig. 4.3 Human–Robot interfacing device

4.2 Additional Module

4.2.1 Hardware

Hardware necessary to construct glazed panels through human–robot cooperation can be largely divided into three groups as follows:

1. Robot controller to control motion
2. End-effector to grip glazed panels into the robot
3. Other devices necessary for construction work

First, the robot controller needs to be able to implement DOF for a mobile platform and a 6DOF manipulator. The first robot controller (Human–Robot Interfacing device) is shown in Fig. 4.3. As seen in this figure, if an operator puts external force containing an operation command on a handler of the robot controller, it is converted into a control signal to operate the robot with ‘sensor A’ (6DOF force/torque sensor; ATI Industrial Automation, Inc.). Here, if the robot comes in contact with an external object, information on the contact force is transmitted to the robot controller through ‘sensor B’ (6DOF force/torque sensor). It is important to note that external force transmitted through sensor B and that transmitted to sensor A should operate separately from each other. In addition, the switch attached to the HRI device should be able to control the manipulator and mobile platform separately. That is, it plays a role of determining whether external force being inputted is a control signal for the manipulator or that for the mobile platform. In this prototype, the operator can select between two communication

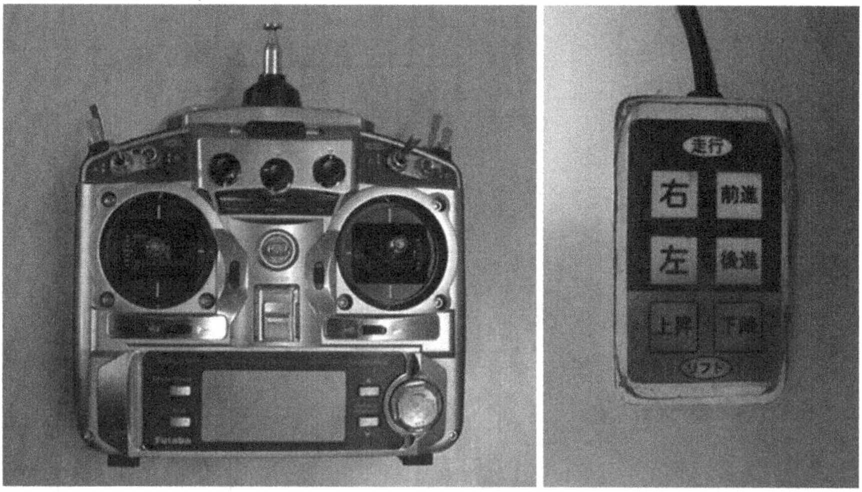

Fig. 4.4 Other robot controllers; *left* wireless controller, *right* teach pendant

methods: wired/wireless control. The wireless control system is used to carry materials long distances or to move a robot to places that are difficult for an operator to reach. The wired control system is used in an emergency. For the wireless communication system, it is then possible to choose between the mobile platform control system and the manipulator control system. In other words, it is possible to control a mobile platform and a manipulator with one wireless controller (Fig. 4.4 left). Each control signal is transmitted to the controller of a manipulator and a mobile platform through a main controller via a RF communication module and a converter. For the wired communication system, it is again possible to choose between the cooperation-based control system and the emergency control system. Unlike the wireless communication system, the wired communication system uses a separate control unit. The cooperation-based control system operates through main controllers including industrial computers and sensors, and the first robot controller (HRI device). The emergency control system can operate through the teach pendant of a manipulator and a mobile platform in emergency situations, as seen in Fig. 4.4.

The end-effector of a construction robot varies according to the properties of the construction materials. Since this study aims at installing construction materials with relatively smooth surfaces, such as curtain wall and glazed panels, a vacuum pad is used as the end-effector. If a vacuum pad located between the HRI device and object makes the contact surface vacuous through a DC motor, the load and end-effector are strongly attached to each other.

Finally, an outrigger to prevent a robot from tumbling, additional safety devices for the operator, and an alarm device to alert neighboring operators of robot operation are necessary, with consideration for the operation environments and characteristics of construction sites. Several controllers and interfaces to

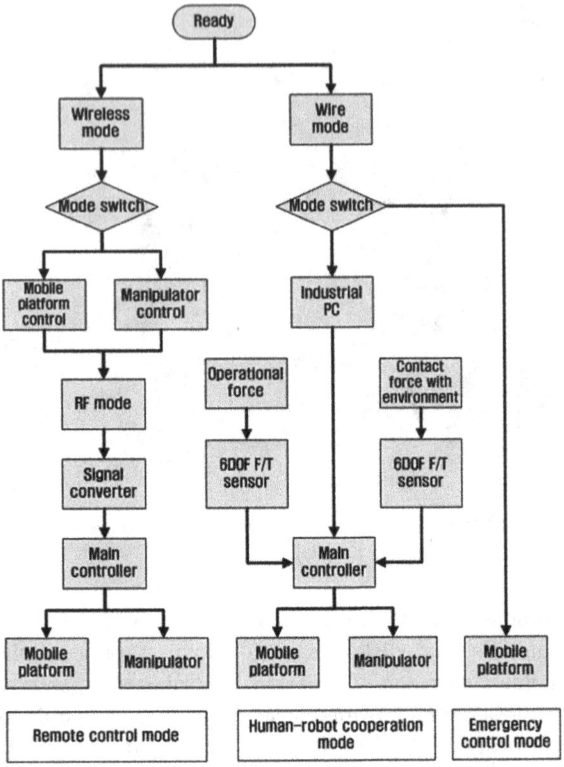

Fig. 4.5 Flow-chart related to robot control methods

implement software and separate power supply systems to operate the robot are also necessary.

4.2.2 Software

The software of the suggested additional module for construction refers to a control algorithm, primarily necessary for installing construction materials by human–robot cooperation. In this prototype, remote control, human–robot cooperation-based control, and emergency control are proposed as methods to control a robot system. Figure 4.5 shows an overall flow-chart for the suggested robot control methods. The operator can select between two communication methods: wired or wireless control. The wireless control system is used to carry materials long distances or to move a robot to places that are difficult for an operator to reach. The wired control system is used to install construction materials by cooperation or in an emergency.

Installing of glazed panels by the human–robot cooperation-based control system can be largely divided as below.

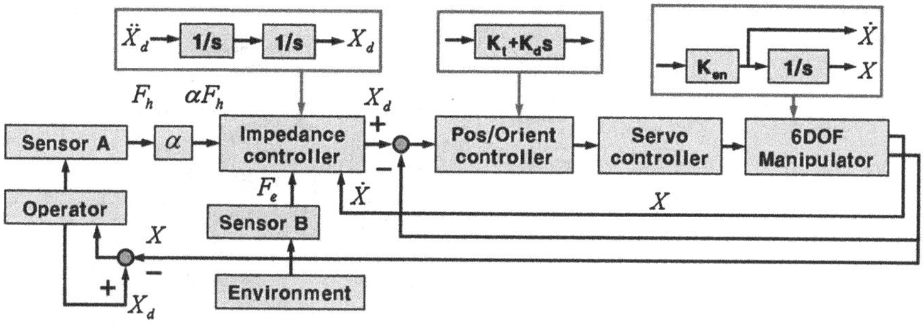

Fig. 4.6 Block diagram of human–robot cooperation-based control

1. Process of transporting panels to an installing site
2. Process of inserting them into the correct position or doing press fits, depending on the environment

In Chap. 2, the former is defined as free space motion (the operation in unconstrained case) and the latter as the operation in constrained case. Free space motion needs rapid movement with relatively low precision while the operation in constrained case needs precise motion with relatively low motion velocity. According to modeling of the interactions among the operator, robot and environment, we designed an impedance controller for the human–robot cooperation (Fig. 4.6). When an operator judges that the position ($\underline{\mathbf{X}}$) to which a robot carries materials fails to agree with the position ($\underline{\mathbf{X}}_\mathbf{d}$) to which he or she wants to carry them, his or her force is transmitted to sensor A. In particular, external force ($\underline{\mathbf{F}}_\mathbf{h}$) measured by sensor A can be used by operators from various age groups through the force augmentation ratio (α). That is, all people, regardless of muscular strength, can operate a robot by the force augmentation ratio. In terms of an operator's inputted force and the contact force ($\underline{\mathbf{F}}_\mathbf{e}$) with environments inputted from sensor B, the target dynamics needed for operation are determined by the following Eq. 4.13 for impedance. Of the dynamics values, the deviation between the target position ($\underline{\mathbf{X}}_\mathbf{d}$) and the present position ($\underline{\mathbf{X}}$) decreases as feedback is received through the encoder of a position/direction controller, resulting in 0. In other words, the current deviation is inputted into a servo controller, which causes a manipulator to pursue the target position value. In addition, it is possible to adapt the operation properties of a robot's motion characteristics by controlling the impedance parameters ($\mathbf{M}_\mathbf{t}$, $\mathbf{B}_\mathbf{t}$) in Eq. 4.13. Relatively rapid and precise motions can be implemented by controlling these parameters.

$$\underline{\ddot{\mathbf{X}}}_\mathbf{d} = (\mathbf{M}_\mathbf{t})^{-1}\{(\alpha\underline{\mathbf{F}}_\mathbf{h} - \underline{\mathbf{F}}_\mathbf{e}) - \mathbf{B}_\mathbf{t}\underline{\dot{\mathbf{X}}}_\mathbf{d}\} \tag{4.13}$$

Here $\underline{\ddot{\mathbf{X}}}_\mathbf{d}$ is the acceleration related target dynamics, $\underline{\dot{\mathbf{X}}}_\mathbf{d}$ is the velocity related target dynamics, $\mathbf{M}_\mathbf{t}$ is the inertia matrix related impedance parameter, and $\mathbf{B}_\mathbf{t}$ is the damping matrix related impedance parameter.

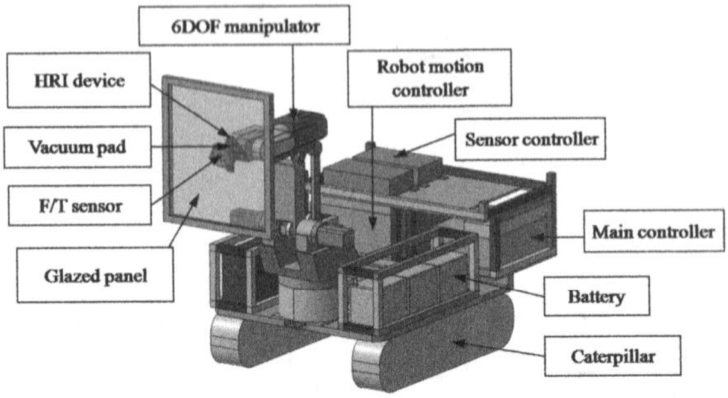

Fig. 4.7 Configuration of prototype system

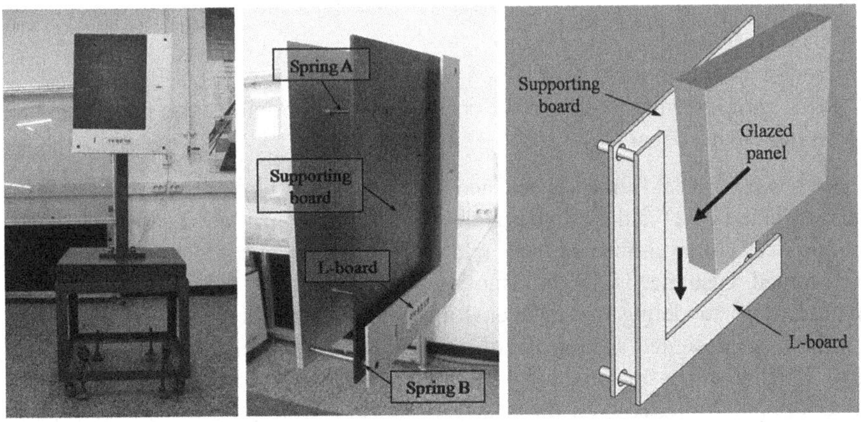

Fig. 4.8 Experimental system and installing method

Figure 4.7 shows the configuration of a prototype system for human–robot cooperative manipulation at construction sites. In this figure, the basic system consists of a 6DOF manipulator and a mobile platform with caterpillar tread; the portion that excludes construction material is an additional module (robot controller, vacuum pad, F/T sensor and controller etc.) for construction works [1].

4.3 Experiments and Results

A simulation for installing glazed panels is implemented to evaluate the performance of the prototype system. The test is implemented indoors with an operation environment similar to that of an actual construction site. An experimental system to implement press pits after inserting construction materials into the correct position was designed as in Fig. 4.8. Inserting glazed panels between the supporting board and the L-board is substituted for actual installation operation. As the gap is narrower than the thickness of glazed panels, they are moved

Table 4.4 Specifications of the glazed panel

Specification	Value
Length	450 mm
Breadth	350 mm
Thickness	20 mm
Weight	50 N
Material	Glass and aluminum

horizontally and vertically with the supporting board connected to 'spring A' being pressed in order to complete the installation operation.

If the supporting board is pressed, it means that contact force occurred; if the length of compression exceeds a certain range (i.e., critical compliance), the result is contact force which causes the robot to move in the opposite direction. In this experiment, a glazed panel was limited to 60 N and below (as shown in Table 4.4), considering the payload of the manipulator.

Figure 4.9 shows a simulation for installing glazed panels through an experimental system. Once a glazed panel is completely gripped to a robot through a vacuum pad at a loading site, the robot is moved relatively rapidly to the vicinity of the installation site through a wireless controller. Precise positioning is performed by human–robot cooperation with a HRI device as Fig. 4.10. In installing glazed panels, an operator is encouraged to collect information on the operation in real time in order to cope with changing environments. Here, the speed or efficiency of operation is proportional to an operator's proficiency.

Figures 4.11, 4.12 and 4.13 show the simulation result of a glazed panel installation using an experimental system. A comparison was made between the contact force ($\mathbf{F_e}$) with environments and an operator's force ($\mathbf{F_h}$), measured by a sensor during the installation of gazed panels. $\mathbf{F_h}$ and $\mathbf{F_e}$ refer to the mean value of forces measured in the x, y, and z directions by a force/torque sensor during operation time t_h and t_e, respectively, as shown in the following Eq. 4.14.

$$
\begin{aligned}
F_h &= \int_0^{T_h} \frac{\sqrt{F_{hx}^2 + F_{hy}^2 + F_{hz}^2}}{t_h} dt \\
F_e &= \int_0^{T_e} \frac{\sqrt{F_{ex}^2 + F_{ey}^2 + F_{ez}^2}}{t_e} dt
\end{aligned}
\tag{4.14}
$$

A total of about 17 s is spent on the simulation, with an average 7 N or less required of an operator to install a glazed panel with human–robot cooperation. In Fig. 4.11, each section can be described as follows:

1. Section A: A glazed panel is carried to an installation position by the operators' force (F_h). As seen in the graph, about 50 N of a glazed panel is carried by about 7 N of force supplied by an operator. The force augmentation ratio is about 7, which is necessary to access the supporting board of the experimental system.

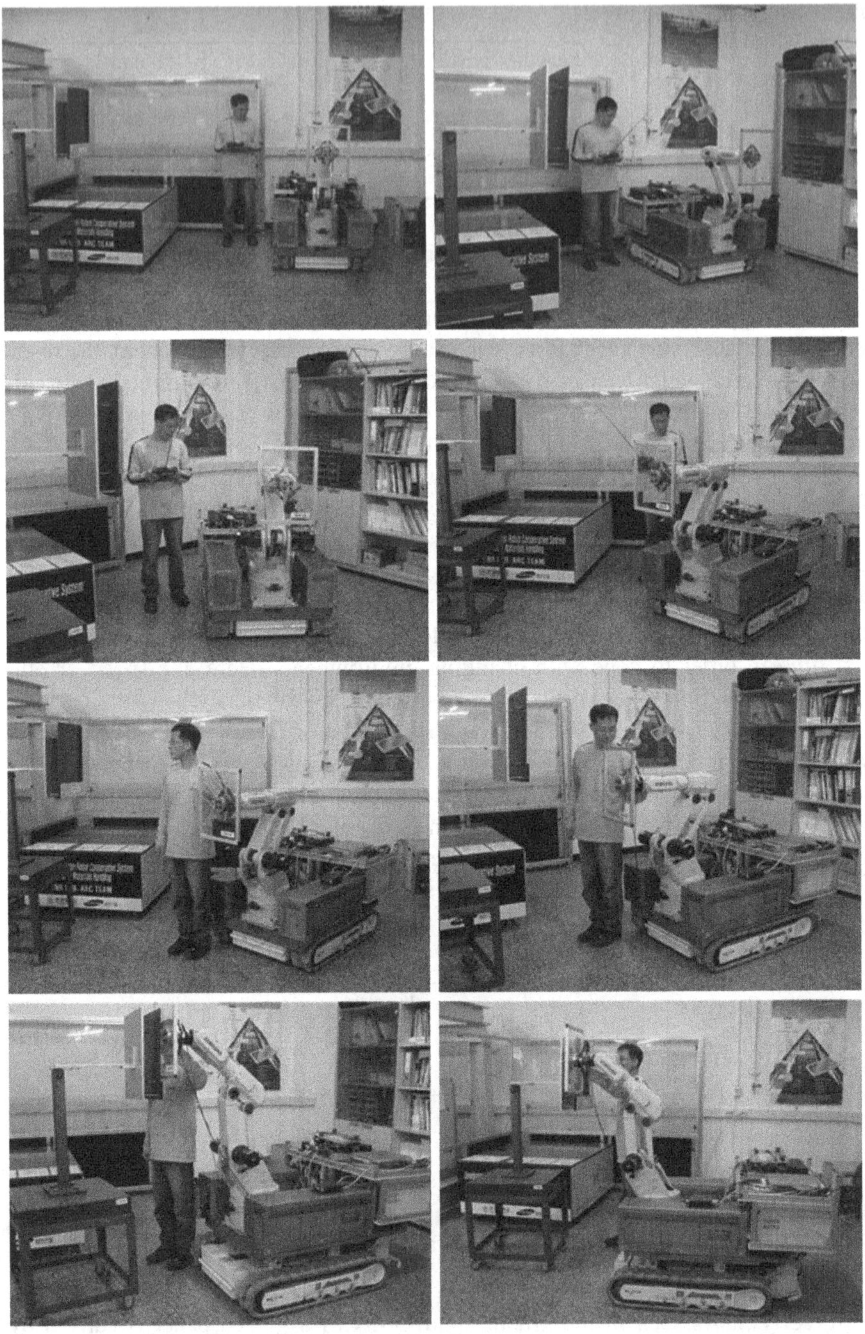

Fig. 4.9 Simulation for installing glazed panels

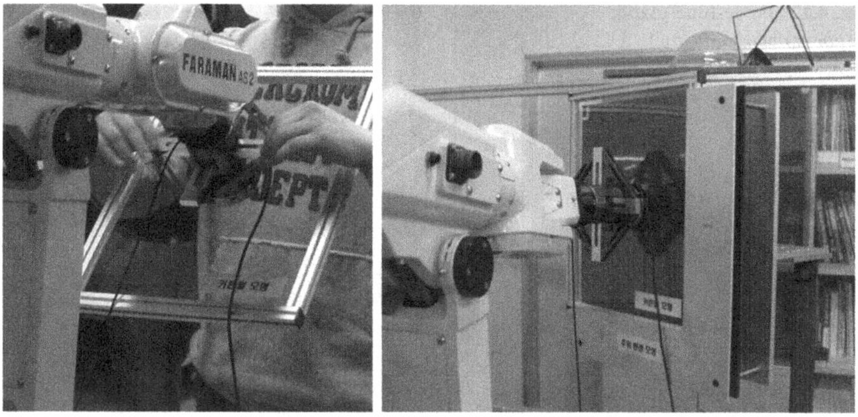

Fig. 4.10 Human–robot cooperation with a HRI device

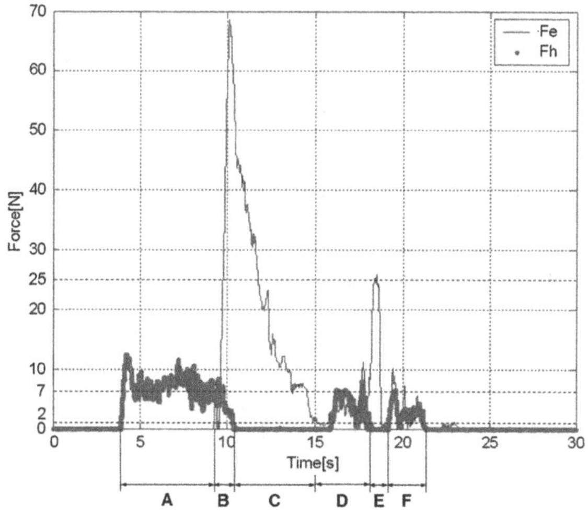

Fig. 4.11 Simulation result of a glazed panel installation

2. Section B: Contact with the environment (experimental system) begins to occur, generating a maximum of 70 N of contact force (F_e). Even at the moment of contact, the operator's force is maintained to press the supporting board of the experimental system.
3. Section C: Not the operator's force but rather his or her torque is transmitted to improve posture. In the end of section, about 2 N of force is generated by the correlation between the external force provided and the impedance parameters of the experimental system. This value is used to press a spring connected to a supporting board into a position with compliance.
4. Section D: A glazed panel is carried horizontally to be inserted between the supporting board and the L-board.

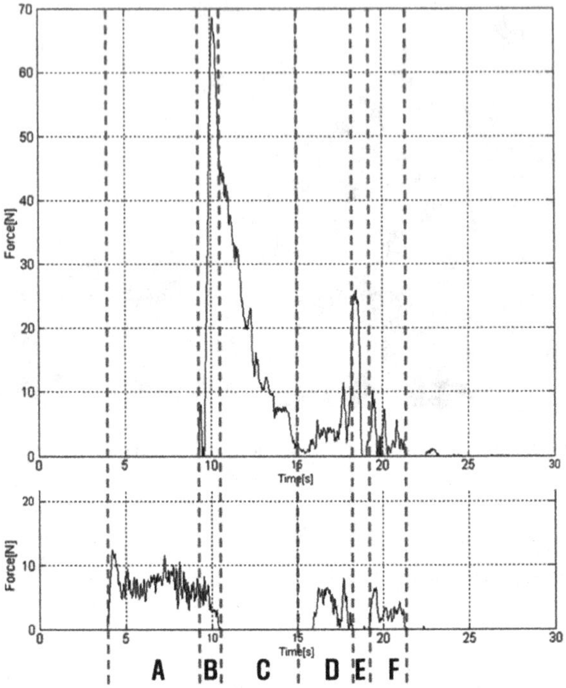

Fig. 4.12 Force-time graph; F_e(*above*) and F_h(*below*)

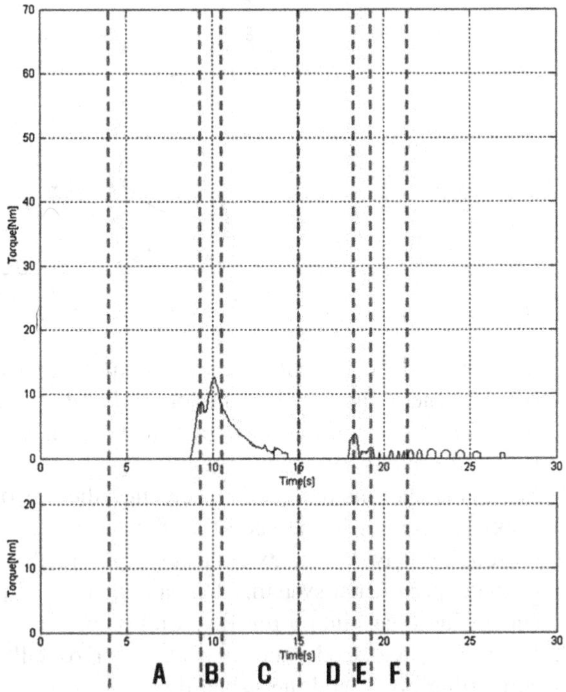

Fig. 4.13 Torque-time graph; T_e(*above*) and T_h(*below*)

5. Section E: A glazed panel is inserted; about 7 N of force is provided by an operator to make press pits, generating about 25 N of contact force.
6. Section F: Inserted horizontally, a glazed panel is then inserted vertically.

Reference

1. S.Y. Lee, Y.S. Lee, B.S. Park, et al. (2007) MFR (Multipurpose Field Robot) for installing construction materials. Autonomous Robots, 22(3):265–280.

Chapter 5
Glazed Ceiling Panel Construction Robot

Abstract The glazed ceiling panel construction robot presented in this chapter combines an aerial work platform and a multi-DOF manipulator. One of the advantages of the proposed robot is the glazed ceiling panel installing by human–robot cooperative manipulation. A HRI device and the vacuum suction device combined with the multi-DOF manipulator are included in this work. Designing of a robotized construction process and field test using the robot is applied on a construction site.

Keywords Vacuum suction device · Robotized construction process · Telescopic boom · Deck · Work-related musculo-skeletal disorders · Field test · Soffit · Manual construction process

5.1 Basic System

As shown in Fig. 5.1, the glazed ceiling panel construction robot is composed of two main parts; a basic system and additional modules. A basic system is also composed of series-type multi-DOF manipulator and an aerial work platform similar to prototype system. However, it is possible to change the elements of not only the basic system also the additional modules according to load specification of building materials.

S. Lee, *Glazed Panel Construction with Human–Robot Cooperation*,
SpringerBriefs in Computer Science, DOI: 10.1007/978-1-4614-1418-6_5,

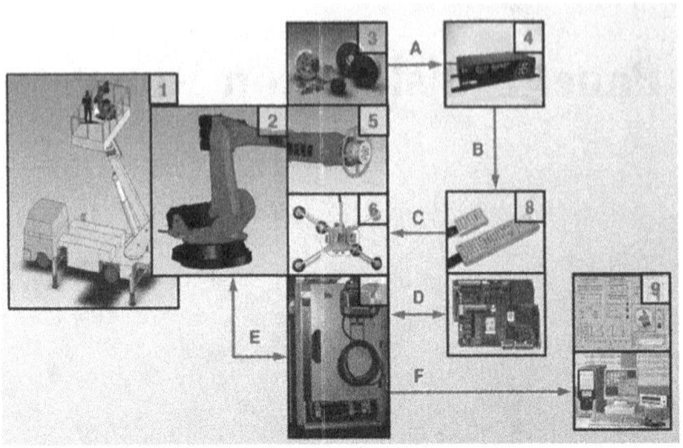

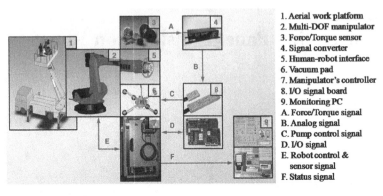

Fig. 5.1 Configuration of glazed ceiling panels construction robot

Fig. 5.2 Multi-DOF manipulator (KUKA Roboter GmbH, KR200 model)

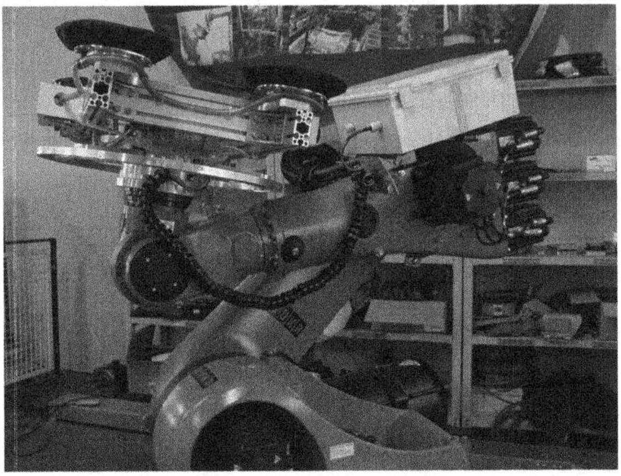

5.1.1 Multi-DOF Manipulator

A multi-DOF manipulator is needed to install glazed ceiling panels, thereby replacing a large amount of human labor, by correlating the operator and manipulator. The manipulator is chosen to help the operators, not to replace them. The manipulator has to be chosen according to the work space and payload. The payload and the weight of any additional devices (vacuum suction device, HRI device etc.) required for installation must be considered. Figure 5.2 shows the selected model (KUKA Roboter GmbH, KR200 model). Figure 5.3 and Table 5.1 show the coordinate systems and D-H Table for the selected model, respectively. In order to control the motion of the manipulator, kinematic and dynamic analysis is required. As operator's safety is influenced by these types of motion, while any singularities in the hardware should be considered carefully.

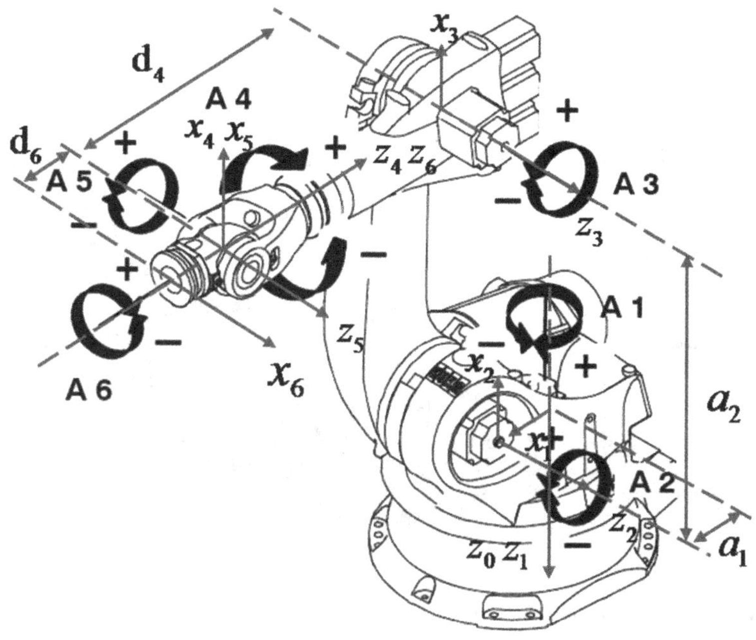

Fig. 5.3 Coordinate systems for the selected model

Table 5.1 D-H Table for the selected model

i	α_{i-1}	a_{i-1}	d_i	θ_i
1	0	0	0	θ_1
2	$\pi/2$	a_1	0	θ_2
3	0	a_2	0	θ_3
4	$\pi/2$	0	$-d_4$	θ_4
5	$-\pi/2$	0	0	θ_5
6	$\pi/2$	0	$-d_6$	θ_6

5.1.2 Aerial Work Platform

Aerial work platforms are designed for enabling altitude work. In this study, the aerial work platform raises the selected manipulator, a glazed ceiling panel, and an operator up to 15 m. A discussion follows concerning the selection of the aerial work platform and the design of the work deck. In selecting a suitable work platform, diverse aspects were considered including mobility, reachable distance, and payload. The work platform must have adjustable movement within a constantly changing work environment. Therefore, considering mobility, a wheel type of work platform was selected, which is mounted on the truck with a telescopic boom. Considering the reachable distance and payload, it is necessary to expand the selection criteria to include not only specific properties but also safety

Fig. 5.4 Selected aerial work platform (truck mounted type)

Table 5.2 Specifications of the selected aerial work platform

Specification	Value
Max. payload	2,000 kg
Rotation range (*Y* axis)	$-17 \sim 70°$
Rotation range (*Z* axis)	360°
Max. working radius	$5 \sim 7$ m
Outrigger	H type
Width (set outrigger)	4.2 m

Table 5.3 D-H Table of the selected aerial work platform

i	α_{i-1}	a_{i-1}	d_i	θ_i
1	0	0	0	θ_1
2	$\pi/2$	0	0	θ_2
3	$-\pi/2$	0	$l_1 + d_3$	0
4	$\pi/2$	0	0	θ_4
5	$-\pi/2$	l_2	l_3	θ_5

concerns. Figure 5.4 shows the selected a work platform that can lift payload of 2,000 kg, and Tables 5.2 and 5.3 show the specifications and D-H Table of the selected aerial work platform, respectively. The rotation range is related to a base coordinate system {O} in Fig. 5.5. In order to implement automated lifting, the kinematic analysis of the aerial work platform (RRPRR type manipulator) must be considered, as shown in Fig. 5.5 and Table 5.4.

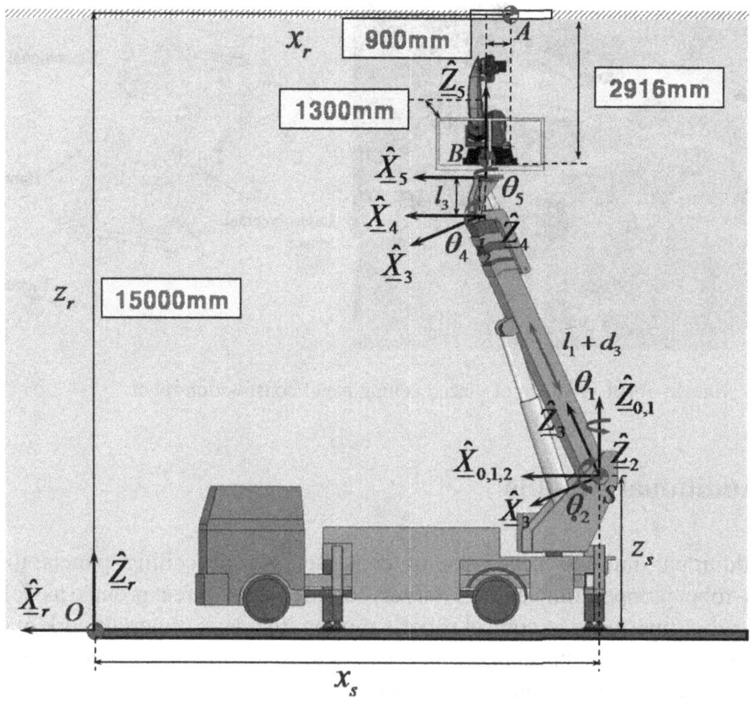

Fig. 5.5 Coordinate systems of the aerial work platform

Table 5.4 Robotized construction plan about each panel number

Panel number	Base point and direction of aerial work platform; *X*, *Y*, direction, (mm)	Base point of deck; *X*, *Y*, (mm)	Rotated boom angle (°)	Extended boom length (mm)
N1-1	−6280, 3000, East	−1450, 1575	−106	187
N1-2	−6280, 3000, East	−1450, 3000	−90	0
N1-3	−6280, 3000, East	−1450, 4500	−73	206
N1-4	−6280, 7500, East	−1450, 6000	−107	206
N1-5	−6280, 7500, East	−1450, 7500	−90	0
N1-6	−6280, 7500, East	−1450, 9000	−73	206
N1-7	−6560, 12000, West	−1450, 10500	74	451
N1-8	−6560, 12000, West	−1450, 12000	90	254
N1-9	−6560, 12000, West	−1450, 13500	106	451
N1-10	−6560, 16500, West	−1450, 15000	74	451
N1-11	−6560, 16500, West	−1450, 16500	90	254
N1-12	−6560, 16500, West	−1450, 18000	106	451
N1-13	−6560, 21000, East	−1450, 19500	−106	451
N1-14	−6560, 21000, East	−1450, 21000	−90	254
N1-15	−6560, 21000, East	−1450, 22500	−74	451
⋮	⋮	⋮	⋮	⋮

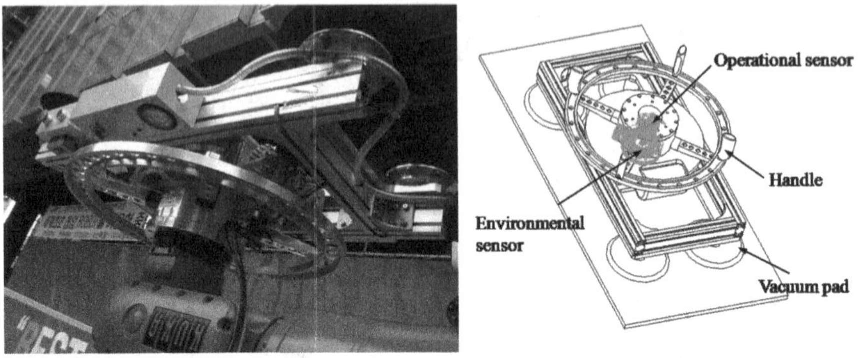

Fig. 5.6 Human–robot interface of glazed ceiling panel construction robot

5.2 Additional Module

The additional module necessary to construct glazed ceiling panels through human–robot cooperation can be largely divided into three groups as follows: human robot interfaces to control robot's motion, the deck of aerial work platform to establish the workspace of an operator and a manipulator, and the vacuum suction device to grip glazed panels on the system.

5.2.1 Human–Robot Interface

The human robot interfaces (HRIs) need to be able to implement interaction between an operator and a robotic system. As similar to prototype, concept of the HRI to control a multi-DOF manipulator is as below. If an operator puts the force containing an operational command on a handle of the HRI (Fig. 5.6), it is converted into a control signal to operate the manipulator with operational sensor (6DOF force/torque sensor; ATI Industrial Automation, Inc.). Here, if the manipulator comes in contact with an environment (i.e., panel frame or obstacles), information on the contact force is transmitted to the manipulator controller through environmental sensor (6DOF force/torque sensor). It is important to note that the force transmitted through environmental sensor and that transmitted to operational sensor should operate separately from each other. From suggested HRI device, operators can handle glazed panels with relatively less force and intuitive operation, while complying with the operator's intention. Also, it allows operators to promptly respond to work environments, changing in real time, through the force reflection in the environmental contacting conditions especially during operations such as press-fit work or peg-in-hole.

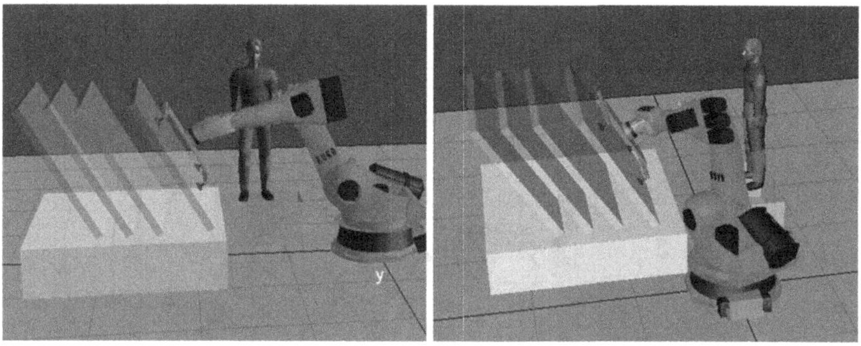

Fig. 5.7 Simulations of operator's motion and manipulator's operation (*Left* case 1, *Right* case 2)

5.2.2 Deck Design of Aerial Work Platform

The deck design of the aerial work platform is important to serve as the work space of an operator and a manipulator, in order to increase productivity and safety. When glazed ceiling panels, an operator, and the multi-DOF manipulator are lifted by the aerial work platform, the position of glazed panels and the manipulator on the deck influences the operator's allowable workspace. Moreover, the deck shape must be optimized to increase operation dexterity, in order to avoid obstacles and allow for a smooth approach to the target position. Two cases were proposed concerning the position of the glazed panels on the deck, as shown in Fig. 5.7. After simulating the operator's motions and manipulator's operation, it was found that case 2 requires more motion by the manipulator than case 1, and needs a larger work space than case 1. Therefore, case 1 was selected, in which the glazed ceiling panel is positioned in front of the manipulator on the deck, as shown in Fig. 5.8. The operator's body size (e.g., arm's reachable length etc.) was considered, in order to prevent WMSD (Work-related musculo-skeletal disorders), as shown in Fig. 5.9. Figure 5.10 shows the configuration of human–robot cooperative system to construct glazed ceiling panels on the designed deck of aerial work platform.

5.2.3 Vacuum Suction Device

The end-effector of a construction robot varies according to the properties of the building materials. Since this study aims at constructing building materials with relatively smooth surfaces, such as glass panels, a vacuum pad is used as the end-effector. The design of a vacuum suction device includes the position of the vacuum suction pads, available payload, and the weight of itself. A position of the vacuum pads should be designed in accordance with the width of the smallest glass ceiling. Four vacuum pads were used as shown in Fig. 5.11, in which each pad can

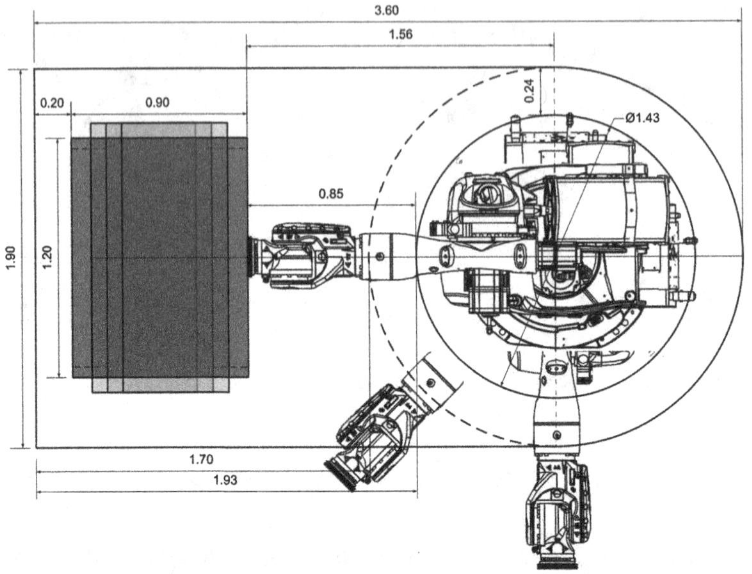

Fig. 5.8 Deck design of aerial work platform based on simulation of human–robot cooperation

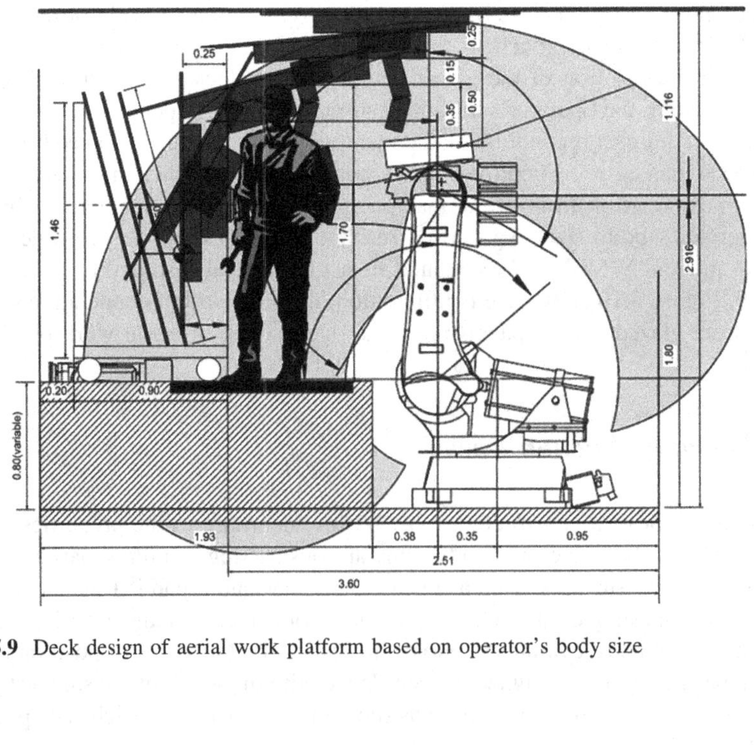

Fig. 5.9 Deck design of aerial work platform based on operator's body size

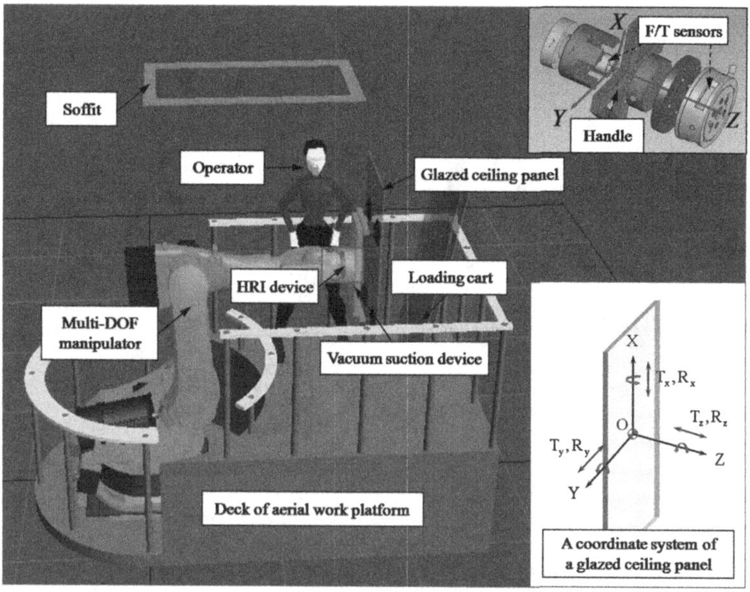

Fig. 5.10 Deck design of aerial work platform based on operator's body size

Fig. 5.11 Vacuum suction device of glazed ceiling panel construction robot

support up to 60 kg. The vacuum pads are positioned in a quadrilateral formation so that an operator can easily put the device on the center of a glazed ceiling panel.

5.3 Design of Robotized Construction Process

On construction site, there exist factors to prevent the equipment from being arranged properly. The construction procedure of installing glazed ceiling panels with the robot can be classified into two processes: the first one is to deploy an

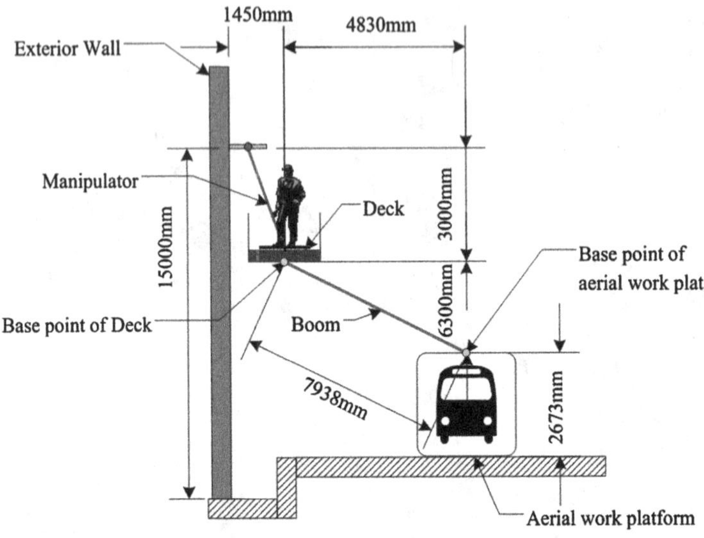

Fig. 5.12 Kinematic analysis of the aerial work platform with deck (*front view*)

aerial work platform on which a deck is mounted and put the deck in the desired position with telescopic booms. The second one is to follow the optimal path making an operator install the glazed ceiling panel with assistance by the robotic system through the human–robot interface. Once the workspace to be deployed is guaranteed, the kinematical method can provide the appropriate position of the deck as shown in Fig. 5.12, Fig. 5.13, and Fig. 5.14. After the kinematic analysis of the aerial work platform about each panel number, the robotized construction plan can be established as shown in Table 5.4. This construction plan is only related to the deployment of the aerial work platform with a deck. As shown in Table 5.4, the base point of the aerial work platform and the deck is described as a position data on the reference coordinate system. It is also served not only the rotated boom angle also the extended boom length in order to install glazed ceiling panels.

After the deployment of the deck, the manipulator's path planning only relies on the workspace between the end-effector and the frame in which the glazed ceiling panel is installed. In real constructing the glazed ceiling panels, many obstacles (e.g., exterior wall etc.) are existed on construction sites which should be considered to install the panels without contacting with them. Therefore, the optimal trajectory of the manipulator should be generated which increase productivity and safety in glazed ceiling panel construction. As widely known in other applications, a computer simulation can solve this problem easily. Figure 5.15 and Fig. 5.16 show the simulation condition and process respectively which are related to find the optimal trajectory of the manipulator. This trajectory is not the only solution to install the panels, and can be modified from the given working conditions and environments. However, it is obvious that the optimal trajectory should

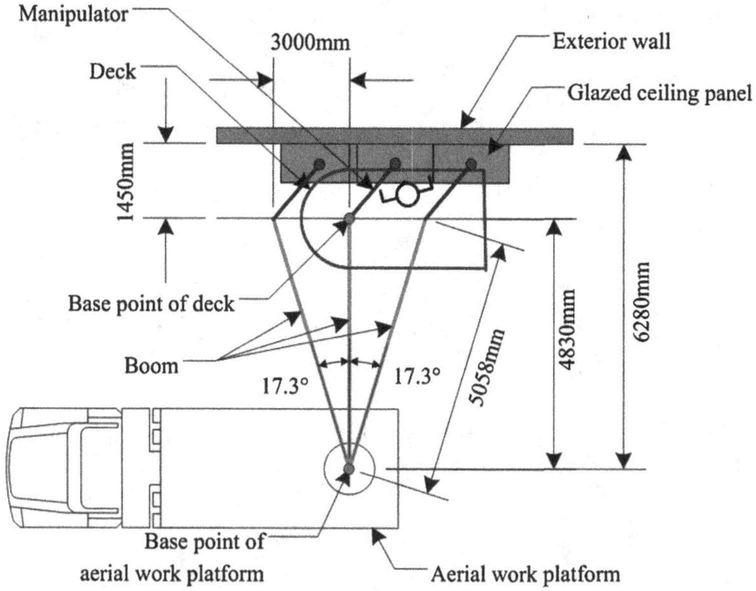

Fig. 5.13 Kinematic analysis of the aerial work platform with deck (*top view*)

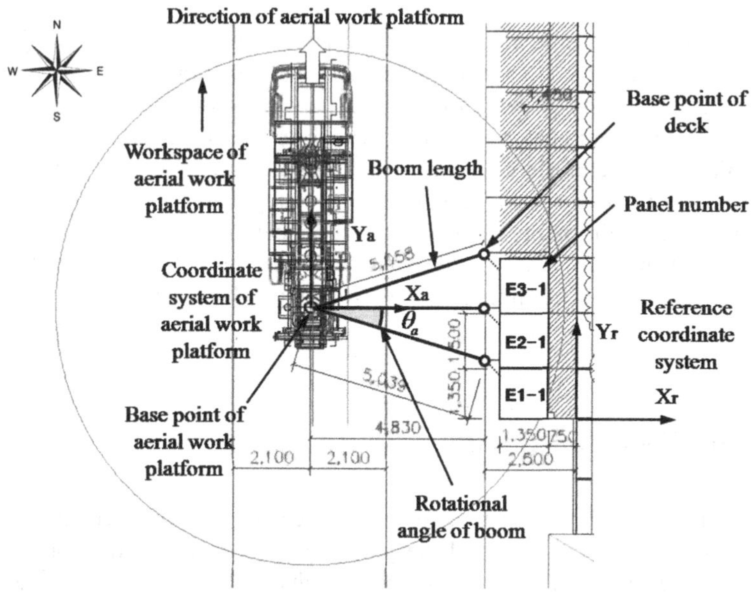

Fig. 5.14 Kinematic analysis of the aerial work platform about each panel number

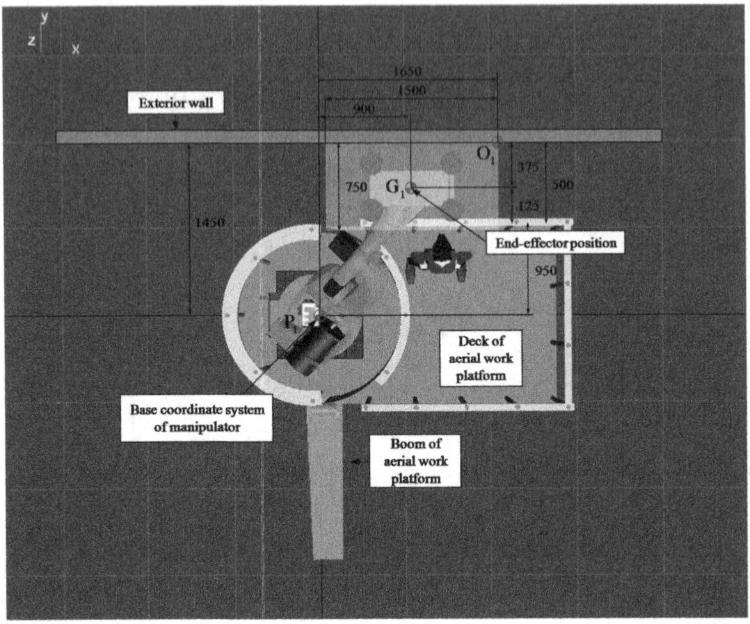

Fig. 5.15 Simulation condition of glazed ceiling panels construction with human–robot cooperation

be considered to increase the productivity and safety in glazed ceiling panel construction, especially in case of human–robot cooperation.

After the simulation of the glazed ceiling panels construction, the end-effector trajectory of the manipulator is deduced which is considered avoiding obstacles and minimizing operator's cooperative motions in Fig. 5.17. It is sure that the motion boundary between an operator and a manipulator is dependent on each other to guarantee the safety of human–robot cooperation. Figure 5.18, Fig. 5.19, and Fig. 5.20 show front view, side view, and top view of the end-effector trajectory respectively according to each axis in the base coordinate system of manipulator.

5.4 Field Tests and Results

Glazed ceiling panel construction robot is developed to construct the soffit, an exterior as shown in Fig. 3.2 and the glazed ceiling panel, an interior in lobby which is to be installed. The total process (including the robotized construction process) of glazed ceiling panel construction can be described as below:

1. The deployment of an aerial work platform on which a multi-DOF manipulator mounted

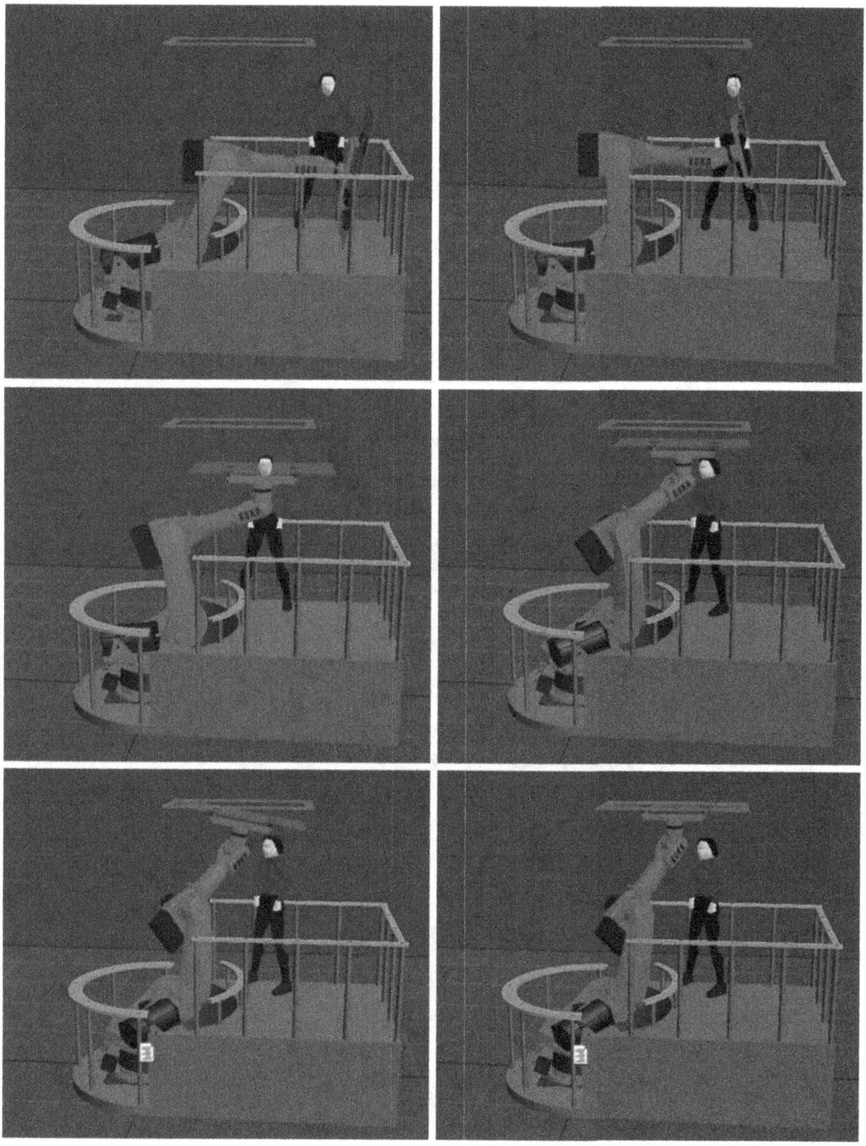

Fig. 5.16 Simulation process of glazed ceiling panels construction with human–robot cooperation

2. Loading glazed ceiling panels on the aerial working platform
3. Approaching the panel near panel frame with human–robot cooperation
4. Installing the panel and finishing the construction work

To prove the proposed construction process and to practice the unfamiliar process, the temporary structure is built to mock the real one though the period of

Fig. 5.17 Simulation result; end-effector trajectory

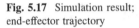

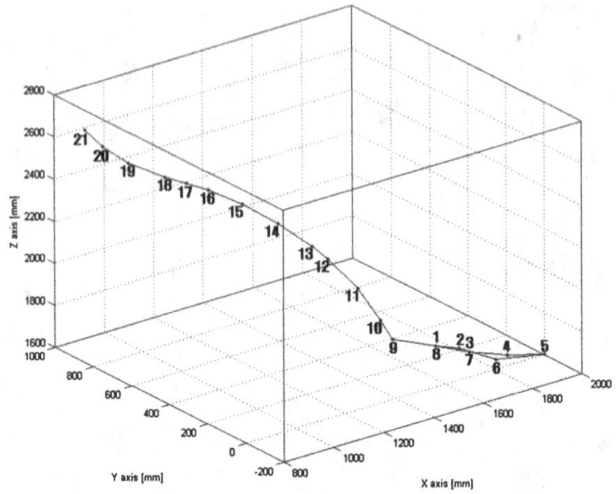

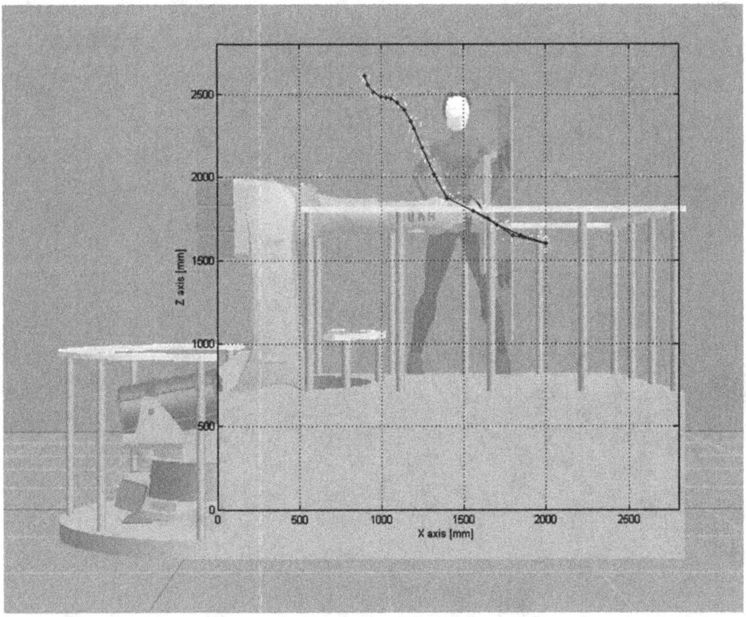

Fig. 5.18 End-effector trajectory of manipulator; *front view*

development limited by the schedule of the construction work, as shown in Fig. 5.21. In this figure, the left shows operational state of the mounted manipulator on an aerial work platform, and the right shows verification process of the robotized construction process on conditions to mock the real structure.

As mentioned in Sect. 5.3, before the construction work on real sites, the construction plan should be made which is contained the detail position of an aerial work platform and a deck corresponding to each panel (numbered in

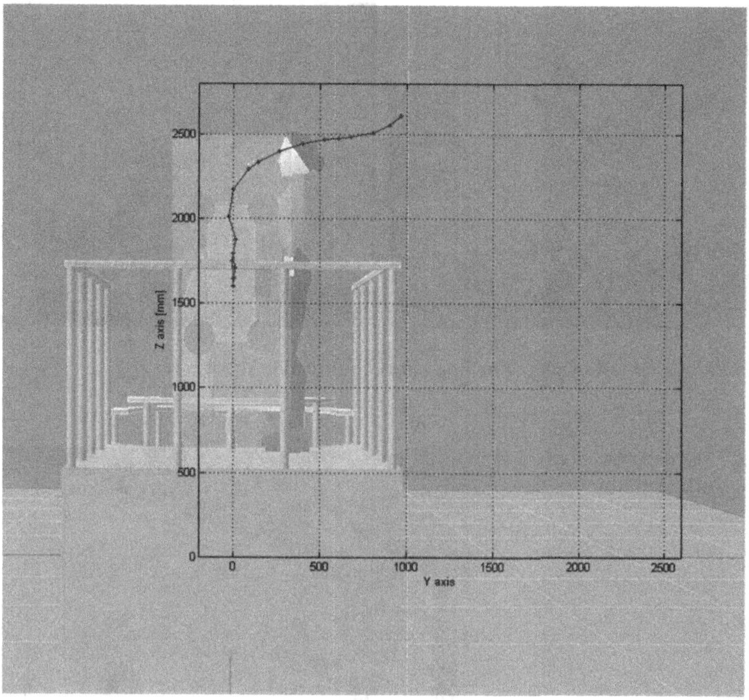

Fig. 5.19 End-effector trajectory of manipulator; *side view*

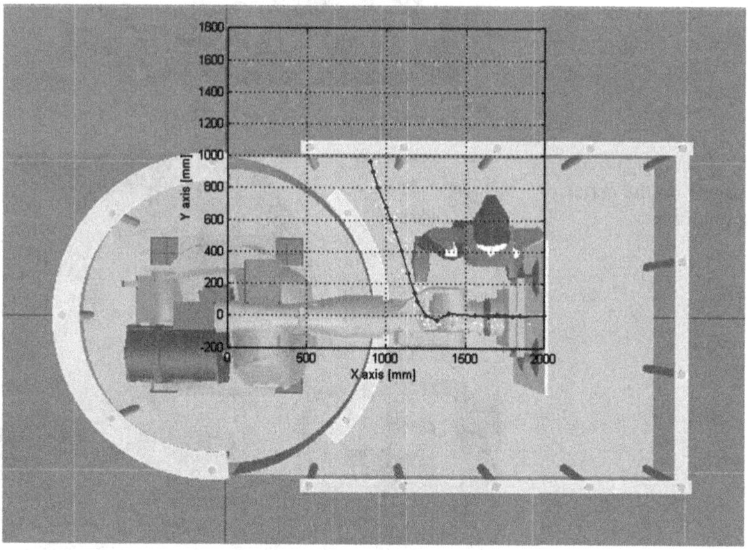

Fig. 5.20 End-effector trajectory of manipulator; *top view*

Fig. 5.21 Mock-up test of glazed ceiling panel construction robot

Fig. 5.22 The deployment of an aerial work platform

Fig. 5.23 Loading glazed ceiling panels on the aerial working platform

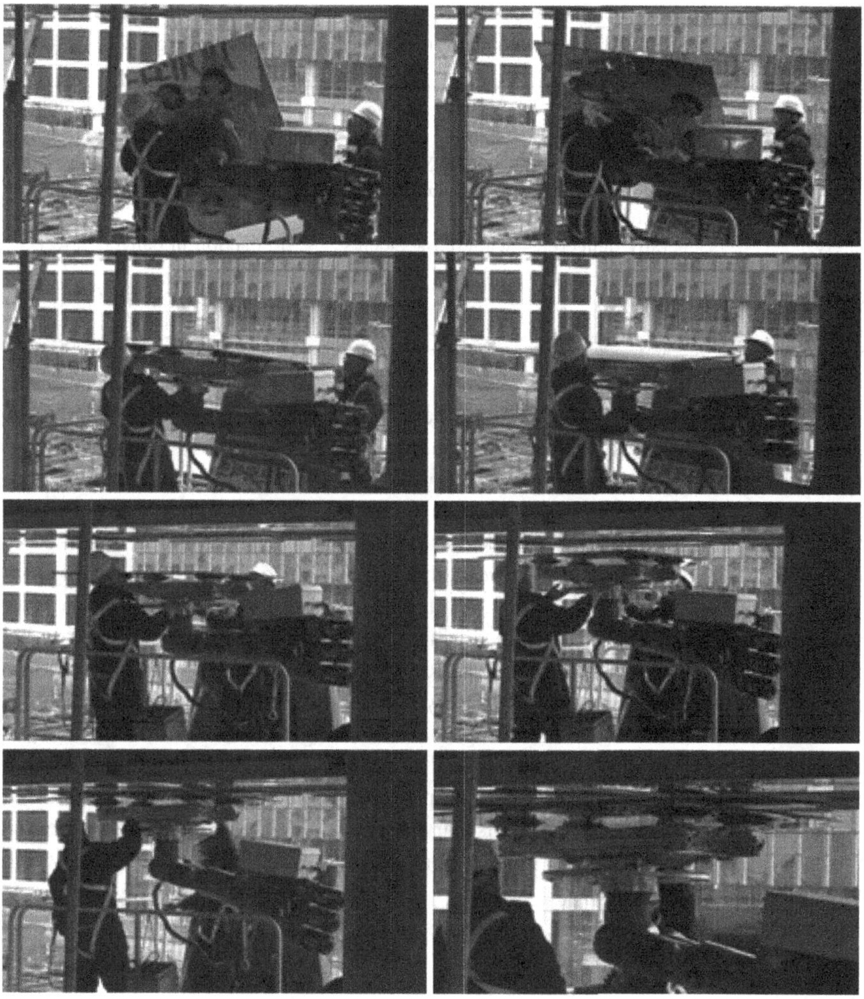

Fig. 5.24 Approaching and installing the glazed ceiling panel

advance). In planning of the construction, it is important that the movement of aerial work platform is minimized since it takes a long time to move the aerial work platform to other positions. According to the proposed construction process, the glazed ceiling panel is installed in panel frame as shown in Fig. 5.22, Fig. 5.23, and Fig. 5.24. Figure 5.22 shows the deployment of an aerial work platform to install the glazed ceiling panel in panel frame. Before lifting the deck, glazed ceiling panels are loaded on the deck. Figure 5.23 shows loading process of the glazed ceiling panel on the deck. The panel is then supported by the aerial work platform and the installation begins according to the robotized construction process shown in Fig. 5.24.

Table 5.5 Comparison of robotized construction method and manual construction method

	Curtain wall construction with manpower and winch [1]	Glazed ceiling panel construction robot
Working time	18 min	Avg. 26 min/piece (including finishing)
Labor intensity	High momentary labor intensity	Generally low labor intensity
Convenience	Profoundly dangerous work under obstacle interference	Generally convenient work
Safety	Generally dangerous; scattered accidents	Reduction in danger; fewer accidents
Number of workers	3	3(deck:2, aerial lift:1)

Table 5.5 shows the results of the field test. Comparison with manual construction process is not executed because the glazed ceiling panel is too heavy to handle by human labor. However, to prove advances in handling heavy construction materials, the existing (manual) construction method of curtain wall construction is introduced [1]. Working time means the whole time consumed in loading the glazed ceiling panel from the ground and installing (including finishing) it in the panel frame. Labor intensity means the degree of manpower strength required of human labors during the glazed ceiling panel construction process. Convenience indicates the degree of difficulty of the installation work, and safety shows derived degree of safety. The resulting comparison and analysis in Table 5.5 can be changed according to the working environment of the selected construction site. In the case of installing a glazed ceiling panel on smaller buildings, the work may depend on manpower. But according to the tendency of current construction trends towards larger and taller buildings, the presented expectations of the glazed ceiling panel construction method and robotic system here provide a bright outlook.

Reference

1. S.Y. Lee, K.Y. Lee, C.S. Han, et al. (2006) A multi degree-of-freedom manipulator for curtain-wall installation. Journal of Field Robotics, 23(5):347–360

Chapter 6
Conclusion and Future Works

The importance of applying the 'Automation System and Robotics in construction' has been raised, as a result of the need for improvement in safety, productivity, quality, and the work environment. Consequently, operations involving automation systems and robots are widely found at construction sites. Since the late 1980s, construction robots have helped operators perform hazardous, tedious, and health-endangering tasks in heavy material handling.

Generally, almost half of construction work is said to be material handling. Materials and equipment used for construction are heavy and bulky for humans. Handling heavy materials has been, for the most part, eliminated for outside work by cranes and other various lifting equipment. Such equipment, however, is not available for inside work. To address curtain walls handling needs for inside work, especially, 'CRCWI (Construction Robot for Curtain Walls Installation)' has been successfully developed. This robot has gone beyond the laboratory and is being applied to actual construction. Through the case studies on constructions, to which CRCWI was applied, however, we could find some factors to be improved. Unlike the automation lines of the general manufacturing industry, construction sites rarely shows repeated operational patterns use to its unstructured processes. That is, construction robots are defined as field robots that execute orders while operating in a dynamic environment where structures, operators, and equipment are constantly changing. Therefore, a guidance or remote-controlled system is the natural way to implement construction robot manipulators. However, during operation of a remote-controlled construction robot, problems arise due to operators receiving limited accurate information; the contact force applied by the robot can damage building materials such as pit from the contact force, thus reducing the ability to respond to the constantly changing operational environments. A human–robot cooperative manipulation, in which an operator can construct materials intuitively, is suggested as a solution.

We developed a robot control algorithm, for installation of bulk building materials in cooperation between an operator and a robot. Especially, considerations

S. Lee, *Glazed Panel Construction with Human–Robot Cooperation*,
SpringerBriefs in Computer Science, DOI: 10.1007/978-1-4614-1418-6_6,

on interactions among operator, robot and environment are applied to design of the robot controller. We examined the influences, which the parameters of the impedance model gave to performance of the cooperation system, through a 2DOF experimental system.

With developing a novel control algorithm, we suggested a new design method of a multipurpose field robot (MFR) that is helped to design heavy and bulk building materials installation robots based on cooperation between an operator and a robot. A MFR can best be conceived in two parts: a 'basic system' consisting of a manipulator and a mobile platform, and an 'additional module' which includes sensor and intelligence technology (including human–robot interface) to execute a particular operation. A MFR have the advantage of executing multiple operations through the use of multiple additional modules in one basic system. Also, continuous system maintenance and improvement is simplified due to the modularization. The development of prototype system of human–robot cooperative manipulation is connected to integrate constituent technology for MFR into human–robot cooperative control algorithm. An installation method for building materials appropriate to the developed system is suggested and a mock installation is carried out.

Until now, we discussed essential technologies of human–robot cooperative manipulation. To apply human–robot cooperative manipulation at real construction sites, we executed additional work required for application. Firstly, according to analysis of job definition and working condition, it is deduced that the conceptual design of a construction robot for installing bulk building materials. Lastly, practical arts (including robotized construction process) for applying to real construction sites are proposed. Finally, after field test at a real construction site, productivity and safety of the developed system are compared with the existing construction method. In this study, we discussed a real construction site to install glazed ceiling panel that is installed 15 m above the ground.

The purpose of this study is to develop human–robot cooperative manipulation technology to solve all kinds of problems generated by the current installation method, which depends on manpower or a low-quality construction robot for the installation of heavy and bulk building materials at construction sites. The essential technologies of human–robot cooperative manipulation are considered through the analysis of an existing installation method. The prototype's hardware design and control algorithm development are achieved using the results of this analysis. The developed prototype system is corrected and complemented based on the results of a performance test. To apply human–robot cooperative manipulation at real construction sites, we executed additional work required for application. After application to real construction sites, evaluation on the productivity and safety of the developed system was done by comparing and analyzing with the existing installation methods.

A manipulator and a mobile platform of the suggested system, are combined to suit various working conditions and building materials as module type. Therefore it is possible to install a variety of *c* building materials in various construction sites. The procedure of installation cannot be defined systematically because it depends on the skill of workers and the construction environment which is unconstrained or not repetitive. However, it is obvious that a human–robot cooperative manipulation is one of solution to address a potential labor shortage and problems in current construction robot technology. To improve technology of human–robot cooperative manipulation, lightweight robot link, robust robot force and position control, flexible robot arm and exoskeletons for human power augmentation will be developed in the future.

Index

A
Aerial work platform, 23, 28, 49

B
Building material(s), 1–3, 23–24, 47, 53, 65

C
Compliance, 7
Conceptual design, 23
Constrained case, 7
Construction industry, 1–2
Construction robot(s), 1–3, 5, 23–27, 29, 37, 47–48, 52–53, 55, 58, 62, 64–67
Curtain wall(s), 1, 3–4, 65

D
Deck, 53–55
Denavit-hartenberg notation, 31

F
Falling accident(s), 23–24
Field test, 47
Force augmentation ratio, 7
Force/torque sensor, 31
Functional requirement(s), 23–24, 26

G
Glazed ceiling panel(s), 23–24, 26–27, 48, 53

H
Human–robot cooperative manipulation, 1
Human robot interface, 23, 52

I
Impedance parameter(s), 7–8, 10–11, 15, 19, 21, 39, 43

L
Labor shortage, 1
Lagrange formulation, 31, 34

M
Manual construction process, 47
Mobile platform, 31, 34–35
Module T&H-bar, 23–24
Multi-DOF manipulator, 23
Musculo-skeletal disorder(s), 23–24, 47, 53

O
Outrigger, 31, 50

P
Panel construction, 1
Press fit, 7

R
Remote-controlled construction robot, 1
Robotized construction process, 47

S. Lee, *Glazed Panel Construction with Human–Robot Cooperation*,
SpringerBriefs in Computer Science, DOI: 10.1007/978-1-4614-1418-6,

S
Soffit, 24, 47
Stiffness parameter, 7

T
Target dynamics, 7
Telescopic boom, 47

U
Unconstrained case, 7

V
Vacuum pad(s), 23, 27–28, 37, 40–41, 53, 55
Vacuum suction device, 53, 55
Vehicle rollover(s), 23–24
Virtual-work theory, 31

W
Wired/wireless control, 31
Work-related musculo-skeletal disorder(s), 47, 53